Women's Education in the United States, 1780–1840

# WOMEN'S EDUCATION IN THE UNITED STATES, 1780–1840

MARGARET A. NASH

WOMEN'S EDUCATION IN THE UNITED STATES, 1780–1840
© Margaret A. Nash, 2005.

First published in 2005 by
PALGRAVE MACMILLAN™
175 Fifth Avenue, New York, N.Y. 10010 and
Houndmills, Basingstoke, Hampshire, England RG21 6XS
Companies and representatives throughout the world.

PALGRAVE MACMILLAN is the global academic imprint of the Palgrave Macmillan division of St. Martin's Press, LLC and of Palgrave Macmillan Ltd. Macmillan® is a registered trademark in the United States, United Kingdom and other countries. Palgrave is a registered trademark in the European Union and other countries.

ISBN 13: 978-1-4039-6938-5    ISBN 10: 1-4039-6938-8

Library of Congress Cataloging-in-Publication Data is available from the Library of Congress.

A catalogue record for this book is available from the British Library.

Design by Newgen Imaging Systems (P) Ltd., Chennai, India.

First edition: March 2005

10 9 8 7 6 5 4 3 2 1

Printed in the United States of America.

Transferred to digital printing in 2008.

# Contents

# Acknowledgments

Archival work depends on good archivists, and I had the advantage of working with some of the best. Staff at the Library Company of Philadelphia, the American Antiquarian Society, Duke University's Special Collections Library, the Southern Historical Collection at the University of North Carolina at Chapel Hill, the Mount Holyoke College Archives, the New York State Historical Society, Cornell University Special Collections Library, and Oberlin College Archives all were of inestimable help and made my work much easier.

I am fortunate to have received funding for my work from various sources. The research phase of this project was supported by grants from the Philadelphia Center for Early American Studies, Duke University Special Collections Library, and the Colonial Dames of America. The initial writing phase was supported by fellowships from the Spencer Foundation, the University of Wisconsin, Madison, and the American Association of University Women. Having the time to revise was made possible by generous support from the University of California, Riverside.

The editors and staff at Palgrave Macmillan have made the process of turning a manuscript into a book as easy as anyone could have made it. Many thanks to Amanda Johnson and Laura Morrison, along with the other folks who worked on this project.

This book began as a dissertation while I was a student at the University of Wisconsin, Madison, and I am grateful for the guidance I received from my committee. Michael Fultz asked tough questions and kept me on the right path, telling me always to "think like a historian." It is because of a class of Michael's that I took during my first semester of graduate school that I became an educational historian at all. Jeanne Boydston went above and beyond the call of duty, reading and commenting on multiple drafts even though I was not her advisee. Both rigorous and encouraging, she pushed me to do work that I wasn't always sure I was capable of; she, however, seemed never to doubt my abilities, and I am grateful to her for believing in me. Bill Reese was consistently supportive and always encouraging of my work, and provided many helpful suggestions for its improvement. Since my moving on from UW, Bill has been an unflagging ally and mentor, and my debt to him is huge.

Many friends and colleagues have proffered support and encouragement. I am deeply appreciative of the environment created by colleagues at the Graduate School of Education at the University of California, Riverside. These colleagues are exactly the people I hoped to be surrounded by when I entered academe: smart, incisive, inquisitive, and generous with their time. My colleagues in the History of Education Society also have been unstinting in their encouragement of my work. Barbara Beatty would not rest until this project became a book, and without her, it might not have happened. Other thanks for mentoring, intellectual challenges, and encouragement go to Linda Eisenmann, Roger Geiger, Jonathan Zimmerman, Kim Tolley, Nancy Beadie, David Labaree, Barbara Finkelstein, and others.

Stephana Ryan generously offered feedback on several chapters, and those chapters are better because of her. Enormous thanks go to Thomas J. Mertz, who read this both as a dissertation and then again as a book manuscript, challenging me in all the ways an academic should, and cheering me on in all the ways a friend should. My mother, Ruth Nash, proofread the entire document and also provided enough homemade baked goods to enable me to survive the process. To Christine Woyshner, Lorene Ludy, Regina Lark, Pat Daly, Barb Bitters, and Susan Harlow: you know what your contributions have meant to the success of this project, and I thank you.

Portions of chapters 2 and 3 were originally published as " 'Cultivating the Powers of Human Beings': Gendered Perspectives on Curricula and Pedagogy in Academies of the New Republic," in *History of Education Quarterly* 41 (Summer 2001): 239–250. Copyright by History of Education Society. Reprinted by permission. A revised version of this material was published as " 'A Triumph of Reason': Female Education in Academies in the New Republic," in *Chartered Schools: Two Hundred Years of Independent Academies in the United States 1725–1925*, by Margaret A. Nash. Copyright 2002. Reproduced by permission of Routledge/Taylor & Francis Books, Inc. Thank you to these publishers for agreeing to the reuse of this material.

Finally, thanks to my parents, Sanford Nash and Ruth Nash, for instilling in me the love of learning.

# Chapter 1

# Introduction

In 1827, Catharine Beecher wrote a widely read treatise on female education in which she made plain her belief in separate spheres of action for women and men. Beecher, daughter of the staunch Calvinist minister Lyman Beecher, was not an advocate of full social or political equality for women. She explicitly stated that "heaven has appointed to one sex the superior, and to the other the subordinate station."[1] Yet, in the realm of formal education, she believed that knowledge "is as valuable to a woman as to a man." She viewed "intellectual cultivation," including the study of ancient languages, as "advantageous to every mind," and in pursuit of academic excellence she modeled her Hartford Female Seminary, founded in 1832, on men's colleges.[2]

Beecher's views encapsulate a central theme—and a central paradox—of the growth of higher education for women in the United States from 1780 to 1840. Institutional growth was fueled by the efforts of educators, philanthropists, and other citizens who believed that women and men had similar intellectual abilities and the same rights to the benefits and pleasures of education. Support for the growth of educational opportunities for women and the increasingly higher academic standards in institutions open to women did not imply a concomitant belief in legal, political, or economic equality. Some writers and reformers linked these issues, to be sure, and either advocated or opposed higher education for women on this basis.[3] But for the most part, a belief in women's capacity for high intellectual attainment did not go hand in hand with a belief in full gender equality. Rather, the realm of the intellect was regarded as being separate from other arenas of life, and was a realm in which few gender distinctions were made in regard to academic studies.

Beecher made her comments in an era in which the United States was both enjoying and suffering from intense growth spurts. Once peace had finally been established after the Revolutionary War and the War of 1812, people turned their attention to more domestic concerns. Factories sprang up in earnest, and cities grew quickly to accommodate the workers in these ever-enlarging worksites. New inventions increased rates of productivity on farms and in more industrial settings, and advances in transportation made it possible to market goods more easily, more quickly, and to farther-flung corners of the country. Early barter economies had long since been replaced by cash economies, and few families could survive on subsistence farming. Waves of immigrants, with distinct cultural backgrounds and traditions, flooded into the country. These changes brought wealth to some, opportunities to many, and challenges to everyone.

Social, economic, and governmental changes provided reasons for increased attention to education. A cash economy, more than a barter economy, required the ability to read, write, and do sums. New types of labor required new types of workers, and the desire to create a national identity required some form of "Americanizing" of youth. Government "of the people, by the people, for the people" also required a citizenry that was informed and could think clearly. Stabilizing a new country, succeeding in an experimental form of government, and integrating immigrants required melding disparate concerns into one overarching interest in national unity. For many national leaders, then, education was high on the agenda of necessary reforms. Thomas Jefferson, Benjamin Rush, and other Revolutionary era heroes all devised plans for increased education. Few of these plans came to fruition in their lifetimes, but the topic was widely discussed from the post-Revolutionary era through the movement to establish public school systems (known as the common school reform movement) in the antebellum era.[4]

That education should seem like an answer to some of these problems is not surprising, as the late eighteenth and early nineteenth centuries were times that placed high value on learning. Beecher's desire for "intellectual cultivation" neatly bridged the Enlightenment philosophies of the late eighteenth century and the quest for "self improvement" of the early nineteenth century. Influenced by both of these movements, combined with an evangelical quest to Christianize the world, Beecher became a prime spokesperson for higher standards of education. In particular, she was widely regarded as a champion of women's education, and she crisscrossed the country establishing schools for women.

Beecher espoused a belief in "separate spheres" for women and men. By this she meant that women should concern themselves with the "private sphere" of home and children, while men should involve themselves in the "public sphere" of paid employment outside the home and in the realms of politics and government. Although she called for a home-based arena of activity for women, she also sought to elevate that sphere. In her most popular book, *A Treatise on Domestic Economy*, a book for housewives that covered everything from recipes to proper plumbing in a kitchen, and from health care to instructions on how to set a table, she repeatedly emphasized the importance of women in maintaining a smoothly running society. Yet she did not *entirely* restrict women to their hearths; she also envisioned a place for them in the classroom. Beecher advocated making teaching a bona fide profession rather than an ad hoc job for unemployed male college students, and she insisted that this newly created profession of teaching should become a profession for women. To Beecher, women who chose the profession of teaching were merely extending their duties in the domestic sphere to include the care and instruction of children in classrooms as well as the children in their own homes.

Beecher was ringed by family members who garnered huge amounts of attention, both in their day and subsequently. Her father was the premier minister of his time, surpassed in ministerial circles only, if at all, by his son, Henry Ward Beecher. Perhaps overshadowing all of them was Henry and Catharine's sister, Harriet Beecher Stowe, the author of *Uncle Tom's Cabin*. Catharine, too, was famous in her own right as an outspoken advocate for women's education, and even more prominently as a writer on domesticity. Given the fame of Beecher and her family, it is not surprising that she has been analyzed thoroughly by historians. Yet, seeing the educational landscape through a Beecher lens—one in which her educational objectives for women are head and shoulders above those of others, and in which an ideology of separate spheres is dominant—skews the view.[5]

Historians frequently group together three brilliant women and the educational institutions they founded: Emma Willard and Troy (New York) Female Seminary, 1821; Catharine Beecher and Hartford (Connecticut) Female Seminary, 1832; and Mary Lyon and Mount Holyoke (Massachusetts) Female Seminary, 1837. Scholars have depicted these institutions as dramatically different from other schools of their time, and credited their founders with inaugurating new types of education for women.[6] *Were* these seminaries so anomalous? If we shift our focus away from these three institutions, what did the rest of the landscape look like? In fact, there were hundreds of other academies and seminaries, both coeducational and single-sex.

In addition to assuming the superiority of Troy, Hartford, and Mount Holyoke, many studies of higher education for women examine academies and seminaries of the early national and antebellum eras largely as precursors to the postbellum college movement. These studies begin with the years following the Civil War when Vassar, Smith, and Wellesley were founded, and use those colleges as a frame of reference for the earlier schools. My approach is radically different. Instead of viewing the academies in the reflected light of their successors, I have sought to comprehend them on their own terms. That is, rather than asking how the seminaries provided a foundation for the women's colleges and the coeducational land grant schools, I have sought to understand the academies and seminaries as institutions in their own right, reflections of the social, cultural, and intellectual mores of their time. What were the impetuses for the growth of higher education in the early national period, and how had these changed by the antebellum era? What were the rationales and justifications for female education? What subjects were offered in these schools, and to what extent was the curriculum determined by gendered presumptions, in particular of marriage and motherhood? How similar or not were men's and women's educational experiences? Most important, what do the answers to these questions reveal about gender in this era?

In this book, I analyze how women's opportunities for higher education progressed from the scattered and short-lived academies of the late eighteenth century to the permanent and highly academic seminaries of the antebellum era. I put the contributions of Troy, Hartford, and Mount Holyoke into a broader context of institutional growth. Doing so reveals a very different picture, one in which thousands of women actively pursued learning, and scores of male and female instructors held high goals for women students. Beliefs that linked a constrained realm of action with limited educational needs seldom held sway, and beliefs in intellectual inferiority only occasionally became obstacles. For most white, middle-class young women, education was a critical component of forging or maintaining a class identity. As the white middle class struggled to define itself, education became a key part of that definition. Gender differentiation within the class was less important than differentiating between the middle class and the socioeconomic classes above and below it. This is not to say that there were no curricular differences for males and females within the white middle class, but that such differences were minor compared to the distance the white middle class wanted to create between itself and people of color and people of other classes.

Prior to the Revolution, some girls and young women enjoyed access to education on par with their male peers. Motivated by religious dictates or a desire to exhibit social superiority, or both, many white women received solid academic education in the eighteenth century. Numerous private schools in the mid-eighteenth century taught academic subjects to girls, and notable religious communities, such as the Moravians in Pennsylvania and North Carolina and Quakers in various places, offered similar academic coursework to both girls and boys.[7] Elite Philadelphians modeled education for their daughters on that of the English gentry, which set out to cultivate reason and rationality in both men and women.[8] Young women in Philadelphia read widely in the arts and sciences, engaged in scientific experiments, conducted historical research, and took part in intellectual conversations.[9] A set of lectures on experimental philosophy advertised the sale of "ladies' tickets," showing an expectation that women would be as interested in, and as prepared to understand, the lectures as would men. Similar newspaper advertisements for displays of microscopes, and for various series of electrical experiments and lectures, welcomed both "Gentlemen and Ladies."[10] Wealthy women and men alike were well educated in a broad array of arts and sciences in the eighteenth century.

This focus on education persisted in the post-Revolutionary era. Indeed, historian Lawrence Cremin has asserted that no topic was as thoroughly discussed in the first decades of the new nation as the need for universal education.[11] Threatened by political and economic instability in the aftermath of the Revolution, some leaders looked to education as the means to cultivate good citizenship, which classic republican theory put at the heart of a stable republic. Influenced by Enlightenment thought, educational reformers encouraged instruction in arts and sciences, not only for their intrinsic value, but also for the purpose of rooting out superstition and prejudice.[12]

From 1780 to 1840, women's opportunities for advanced education burgeoned. In the 1780s, women who sought education beyond the rudiments found it primarily in temporary, short-lived schools (variously called academies, seminaries, or institutes) that were open for only a few weeks or months at a time. The subject matter varied from school to school, but most academies taught the English branches, geography, and arithmetic. By 1840, formal higher education for women was well established. Literally hundreds of institutions existed, many of which operated on a permanent basis. Principals and proprietors raised academic standards continually over the years and offered longer school terms than they had in the 1780s. They subscribed to a systematized course of study in which students took courses in a logical progression, and they expected students to stay at least a year, if not for the entire three- or four-year program. While most female scholars still attended seminaries, academies, or institutes, either coeducational or single-sex, a few attended the small number of colleges that admitted women as well as men. Further, by 1840, the three institutions that have been identified by historians as the premiere female institutions—Troy Female Seminary, Hartford Female Seminary, and Mount Holyoke Female Seminary—had all been founded.

At this point, a comment or two may be in order about the various terms for institutions for higher schooling. Defining exactly what was meant by the terms institute, academy, seminary, and college is a difficult, perhaps impossible, task. There were no clear definitions at the time; indeed, higher education was "a flexible nineteenth-century concept."[13] Certainly, the nomenclature implies very little, if anything, about the curriculum or the depth of study. Generally speaking, a college was a degree-granting institution offering a three- or four-year course of study primarily designed to prepare students for the ministry, law, medicine, or other professions. Until the late antebellum period, virtually all colleges were all-male institutions. Academies also offered education beyond the rudiments. They were more likely to offer practical courses such as surveying and navigation than colleges, but many also offered curricula indistinguishable from the colleges. Thus, in his history of education in the United States, Robert Church discussed male colleges and academies together, using the terms interchangeably, "for in essential features and functions the two institutions were fundamentally similar."[14]

If for male schools the terms "academy" and "college" were virtually interchangeable, for female schools the same was true of "academy" and "seminary." Trustees and proprietors were somewhat more likely to name women's institutions "academies" in the early republic and "seminaries" in the antebellum era.[15] However, few educators, school founders, or students used any one designation consistently, employing various names for their institutions, often even within a single descriptive paragraph. Trustees of Boston's high school for girls variously referred to the same institution as a school, a high school, or a seminary.[16] Trustees of Adams Female Academy began their 1831 brochure by speaking of "this seminary."[17] Similarly, the writer of a catalog for the Arcade Ladies' Institute in Rhode Island sometimes used the term "Institute" and at other times the term "Seminary."[18] The president of the trustees of the Albany Female Academy in a speech in 1834 sardonically summed up the mania for new names: "we do not rely upon the magic of a name." Our institution, he said, "is neither a High School, nor an Institute. It is neither a Polytechny [sic], nor a Gymnasium. We have no disposition to play upon the credulity of the community by the use of new and pompous titles. It is what it was from the beginning, an Academy."[19]

The notable exception to this rule of flexibility in taxonomy was the designation "college." Whether a female school was called a seminary, an academy, or an institute, seldom was it called a college, although there were a few female "collegiate institutes." Indeed, virtually no institution called a college admitted women before the 1830s. In this regard, there was an enormous difference in higher education for men and women. Although elite colleges continued to be all-male, after 1830 less elite colleges began to admit women. In fact, one scholar has identified close to 20 coeducational antebellum degree-granting colleges.[20] Therefore, although most colleges were male only, dozens opened their doors to women after 1830. To the extent that women were denied entrance, however, female exclusion from colleges was a reflection of different social positions and of occupational segregation in the late eighteenth and early nineteenth centuries, and not a reflection of a belief in different intellectual capabilities.

The extension of college education to women by the 1830s was in part a result of new views of the purpose of college education. Popular views on what constituted a "college" changed dramatically between 1780 and the huge collegiate growth spurt of the late antebellum era. In the late eighteenth century, an extremely small proportion of the American population—an estimated two-tenths of one percent—attended college for even one year, and even fewer graduated.[21] In this period, college education was strongly associated with training for specific, male-only occupations, such as medicine, law, and the ministry. By the antebellum era, the proliferation of denominational colleges, combined with an evangelical fervor, reflected a new purpose. No longer training an elite few for particular positions, colleges prepared young people, both men and women, to participate in creating a Christian nation. The purpose of college education had changed, and the new purpose created more room for the inclusion of women.

The lack of clarity in the names of institutions has created confusion in work on the history of higher education in the United States. When historians began looking at higher education for women, they began with institutions that *called* themselves colleges. This almost necessarily excluded antebellum institutions, and resulted in a focus on the elite private women's colleges or the land grant colleges. The first historical work on higher education for women posited a trajectory from nonexistent formal education for colonial women, most of whom remained illiterate, to fledgling academies in the early national era that taught needlework and a smattering of geography, arithmetic, and English grammar, to antebellum seminaries, a few—and only the best—of which struggled to offer a curriculum similar to men's colleges, and finally to the women's colleges and coeducational land grant universities of the postbellum period, in which women finally had access to a full collegiate curriculum.[22]

Because historians touted women's colleges and land grant schools as the first significant instance of intellectual or academic achievement for women, they made two problematic assumptions about the seminaries. First, they tended to view seminaries as vastly inferior academically, both to the later colleges and to the men's schools of their own period.[23] They pointed to a few female institutions (Troy, Hartford, and Mount Holyoke) as exceptional in their era and as being important stepping stones to the later colleges. Because historians had concluded that high academic standards for women were not fully accepted until the late nineteenth or early twentieth centuries, when a plethora of institutions for women proudly adopted the term "college," they assumed that much of the general public disapproved of or opposed the opening of academies and seminaries and that school founders who pushed for higher standards were doing radical and solitary work.[24]

Second, historians have explained the alleged inferiority of female academies by asserting that school founders, teachers, and the students themselves were constrained by the ideology of "separate spheres." Seeing the educational landscape through this prism led to an assertion that there was a strong distinction between male and female education. Either people of the day viewed women as not capable of higher learning or they viewed women's lives as too narrow to need education, or both. Regardless, men's and women's education must have been starkly differentiated; else, how uphold the tenets of the doctrine of separate spheres? Historians have emphasized one strain of thought that held that women were intellectually inferior to men, and therefore have assumed that schools for women could not possibly have been as intellectually rigorous as men's schools. The few schools that were rigorous must have been gargantuan efforts to prove women's intellectual mettle to a disbelieving public.[25]

Not only did the general population view women as intellectually subordinate to men, according to this model, but also that women and men were destined for life in different spheres of activity, and therefore required dramatically different educations. Education, according to some historians, was based on the principle that "women needed to be educated separately and differently from their brothers."[26] According to these historians, not only were women viewed as less intellectually competent, but their different life paths as wives and mothers also led to different educational needs; if men and women were being prepared for very different life roles, the curricula must have been highly gender differentiated.

Both of these assumptions—of a widespread belief in women's intellectual inferiority and of a broad adherence to an ideology of "separate spheres"— are being questioned by historians. Recently, historians have begun to reassess early Americans' views of women's intellectual capabilities, and have recovered points of view more favorable to female education. A vibrant transatlantic stream of Enlightenment thought held that "the mind has no sex."[27] Followers of this ideology, believing that there was no biological difference between male and female brains, supported advanced education for women. Studies of patterns of female reading in the antebellum era reveal women who eagerly sought out images of learned women who demonstrated "intellectual agency" for their role models, apparently with little fear of becoming unmarriageable bluestockings.[28] Close examination of readers' diaries (i.e., diaries of what one read and what one thought about it) and of book selections made by library patrons showed that antebellum men and women chose virtually the same material for leisure reading, countering longstanding scholarship that held that women read "lighter" material, such as novels, while men read more serious work. The hundreds of readers' diaries studied did not reflect a cultural assumption that women were not capable of reading serious news or literature.[29] The research on which my study rests sustains the position that the idea of female intellectual inferiority was not as ubiquitous as the earlier historiography suggested. While some writers promulgated a view of men's and women's different capacities, the dominance of this view has been exaggerated; there were contemporaneous ideologies emphasizing women's capacity for rational thought and higher learning.

The paradigm of "separate spheres," first defined by historians in the 1960s and 1970s, has been challenged and refined in the last decade or so, as well. Although this ideology was present in prescriptive literature—indeed, early nineteenth century essays, articles, and sermons were replete with references to "woman's sphere"—many historians agree that the phrase did not reflect the reality of women's experiences, and that the spheres never were all that separate.[30] For instance, Mary Ryan contends that the ideology of separate spheres that divided the world into public/private and male/female was not an accurate description of actual behavior, but an attempt to create order out of chaos, to impose a system—or the perception of a system—on a disorganized world.[31] This implies that, despite the volume of prescriptive literature advocating domesticity, large numbers of women were not fulfilling its tenets. Proponents of domesticity stepped up their insistence on separate spheres to the same extent that some women refused or were unable, due to class, race, and/or inclination, to remain in their sphere.[32]

The concept of separate spheres depends on clear demarcations between public and private, but scholars have demonstrated how indistinct those lines actually were.[33] What constituted public and private behavior, and therefore whether that behavior was deemed appropriate for women, was in flux in the early national and antebellum eras. Inhabiting the private sphere did not prevent women from voicing their concerns to the government, for instance. Susan Zaeske documents women's participation in the act of petitioning the government even as that act changed discursively from supplication to more radical insistence on rights.[34] Julie Roy Jeffrey reveals the wide array of means by which abolitionist women engaged in the public sphere without partaking in voting, and without holding or running for public office. Unable to enter those segments of the public sphere, women did avail themselves of access to other segments; they wrote essays for newspapers and journals, spoke publicly, sponsored pageants and picnics, and created fairs as marketplaces, "not only for consumer goods but for public opinion, for ideas and ideologies."[35] The lines between "public" and "private" were neither clearly defined nor fixed; instead, they were permeable and constantly being negotiated.[36]

So imprecise are the terms "public" and "private" that we might need either to define them more clearly or find more accurately descriptive ones. Some scholars suggest using the term "social sphere" to describe activities that are neither lodged solely in the home nor are explicitly governmental. This approach illuminates a long list of activities participated in or conducted by both men and women, including civic societies, reading circles, benevolent organizations, and activist agencies. Other scholars suggest defining "public" broadly, so that it includes not only public buildings, such as courthouses, churches, and businesses, but also figurative public spaces, such as newspapers, novels, and journals, as well as the amorphous groups of citizens who create "public opinion." A broader definition of public certainly permits women's lives to figure more prominently, and more accurately, than does a narrow definition.[37]

These challenges to the ideology of separate spheres have barely begun to have an impact on the history of women's education. Much of the literature about the antebellum seminaries is deeply rooted in assumptions about "separate spheres." Assessment of seminaries takes the form of demonstrating either how the schools supported or countered the ideology of separate spheres. Many historians have argued that the seminaries reinforced the ideology by training women for passive and "traditional" family roles, while others have argued that some antebellum leaders, such as Mary Lyon, almost single-handedly challenged female domesticity by urging women to assert themselves.[38] All of these scholars accepted the ideology of separate spheres largely as a fixed entity, and asked only how a particular educator or institution represented acquiescence to or rebellion against domesticity.

My examination of the development and transformation of higher education for women suggests that assumptions regarding women's work as wives and mothers did not result in curricular ideals very different from those held for men. Few academies and seminaries of the early national or antebellum eras taught courses directly related to wife- or motherhood, such as childrearing techniques or domestic management. Rather, most educators believed that a general liberal arts curriculum, similar to that for men, was good preparation for any life role. Students, teachers, and school founders alike largely concurred with Catharine Beecher in believing that a broad-based education was as valuable to women as to men.

The ideology of "separate spheres" has shaped the historiography on women and education in another way, as well. Because most formal learning occurred outside the home, historians may have too quickly positioned education in the public realm. To the extent that these historians also situated women in a "private sphere," they may have assumed women were not welcome in "public" educational arenas. As education became increasingly institutionalized, it did become more public—not in the sense of being entirely funded through municipal or state dollars (although the public did fund schools to differing degrees in different places over time), but in the sense that by the late eighteenth century education had moved out of the private realm of the home. Academies and seminaries held examinations to which they invited the public, and reports of the examinations were printed in local newspapers, which also frequently published student essays, poems, dialogues, and other forms of writing. In these and other ways, institutions, teachers, and students were in the public eye. At the same time, however, young women in many academies and seminaries lived in a protected environment, governed by strict rules covering nearly every aspect of their lives, from receiving visitors to walking into whatever town was nearby. Women were cloistered within seminary walls, so much so that critics sometimes disparagingly referred to seminaries as "Protestant convents." Institutionalized education for women, then, inhabited a place of liminality. Some aspects of formal education existed solidly within the public eye, and some aspects were veiled from public view. This mixed status may explain, in part, why the "public" nature of education was not an impediment to women's attendance. To the extent that education was public, it was so in a form that was acceptable.

The quasi-public, quasi-private condition of education also points to problems with a paradigm based on a public–private dichotomy, and with assigning the sexes to one or the other end of the spectrum. Few places were either entirely private or entirely public, and few places were either entirely female or entirely male. If we continue to see women as relegated to a small private sphere, then we miss a wide range of their realms of action, and we also miss a broader understanding of the meanings, purposes, and uses of women's education.

Women in the early American republic certainly did face constraints. Women and men did not have full legal, social, economic, or political equality. What also is the case, however, is that not all men had the same legal, social, economic, or political rights as each other. It is inaccurate to speak of "men's rights" and "women's rights" as though what was true for any one man or any one woman was true for all. Class and race made enormous differences, and the difference that class, race, and gender made changed over this period. Nor was "woman's sphere" a fixed concept; it did not mean one clear thing across the scope of the nineteenth century, or across geographic regions, or across class lines. Prescriptive literature spoke frequently of "woman's sphere," but the definition of that sphere was constantly in the process of creation and negotiation; or, as one scholar phrased the process, it was "internally dynamic and constantly interacting with the surrounding economy and society."[39] Quite possibly, "woman's sphere" was as nebulous to people in the antebellum era as the phrase "family values" is to us today; there is a political worth to using that phrase as rhetoric, but it is certainly not the case that all Americans agree on exactly what it means.[40] Therefore, using "separate spheres" ideology to explain women's education in this period necessarily limits our understanding both of education and of the construction of gender.[41] Instead, we must examine the processes of producing ideals of gendered social relations.

This study, then, applies these new perspectives in women's history to the study of higher education for women. It examines women's education in the early national era and the antebellum era, looking both at theories and rationales for that education, and at what we can discern about what instructors actually taught.

I begin my narrative immediately after the Revolutionary War. In chapters 2 and 3, I look at the early national period—first at the theories behind women's education, and then at actual practice. In chapter 2, I discuss strains of thought that not only allowed for, but also encouraged, the intellectual growth of women. Although there were writers who put forth a belief in women's limited potential for academic attainment, there were many other writers who argued for higher education for women. I discuss Enlightenment beliefs that education was a mark of civilization, and the impact of Lockean theories of environmentalism on education. I next turn to the ethos of civic republicanism, which posited that education was a way to preserve the republic. I examine the curricular ideals of educational leaders of the period, as well as challenges to the growth of higher education for women. The discourse on advanced education for women in this era reflected both the rhetoric of human rights and Enlightenment ideals about intellectual equality.

In chapter 3, I look closely at academies of the early national era. I compare the curricula for women and men, in both academic and nonacademic subjects. I argue that, due to educators' investment in the Enlightenment beliefs and the ethos of civic republicanism described in chapter 2, along with the reality that more men than women were qualified to teach advanced subjects, both the curricula and the pedagogy were similar for men and women in most academies of the period. Core subjects were virtually identical. Most curricular subjects differed only in the area of courses related to specific vocations.

In chapter 4, I discuss female education and class formation in the antebellum era. Individuals who saw themselves as members of a newly emerging "middling class" struggled for self-definition. Some women and men justified advanced education for women on the basis of a heightened evangelical emphasis on the need to Christianize the nation. In addition, for many who wanted to be part of the "middling classes," education became an emblem of class status. This is evident not only in the enormous growth in institutions, but also in the popularity of an ethic of self-improvement. In this chapter, I also discuss the civic emphasis on maintaining order for the sake of social stability, intertwined as it was with the rapid development of the common school movement. The swelling numbers of common schools created a need for female teachers, who, in turn, needed education. Women of the "middling classes" flocked to teaching as a respectable position that would provide them with some degree of economic self-sufficiency.

In chapter 5, I maintain that a significant reason why women pursued education that has been most overlooked by historians was their sheer longing for learning. Religious convictions and the imperative to be educated enough to save souls, as well as the desire for relative financial security, surely inspired many women to pursue advanced education. In addition, many people in the antebellum era expressed a love of knowledge for its own sake. School founders appealed to this desire in their advertisements for students, and women recorded their hunger for learning in diaries and letters. Because most educators believed that women and men were equally capable of learning—and because women actively sought out schools with increasingly higher academic standards—the curricula and pedagogy in antebellum institutions were very similar for women and men. For as much as advocates for female education discussed how education would make women better wives and mothers, virtually no school taught domestic economy, housewifery, or childrearing.

In chapter 6, I discuss some of the limitations of higher education for women. Virtually all of the students in institutions of higher education in the 1820s and 1830s were white and either middle or upper class. Education became one way the middle class could distinguish themselves both from the very poor and the very wealthy. In fact, higher education for females before 1840 may have been relatively uncontested by middle-class white men because it was an integral part of creating a distinctive white middle-class culture. It may be that before 1840, gender was not as salient a category as were issues of class and race. Finally, I briefly bring the discussion up to the twentieth century. As the numbers of women attending or graduating from seminaries, high schools, and colleges increased in the late nineteenth century, opposition grew. The late nineteenth and early twentieth century saw the coordinate college movement, which had the clear intent of segregating women away from men. One possible explanation, along with the eugenics movement and fears of white "race suicide," may have been the growing woman suffrage movement that explicitly linked women's education with women's political rights. This generation of women was no longer willing to accept intellectual equality without social and political equality.

# Chapter 2

# "Is Not Woman a Human Being?"

## Discourses on Education in the Early National Period

"Let me ask this plain and rational question," wrote the British essayist Anne Randall in 1799, "is not woman a human being?" Her plaintive query reflected Enlightenment thought and sought support for female education based on a belief in women as sharing the essential human quality of rational thought. Central to Enlightenment conceptions was the assertion that the capacity to think rationally set humans apart from animals. In this view, women, as human beings, were capable of abstract reasoning and therefore would benefit from exposure to the arts and sciences. If Nature bequeathed this capacity to women, some argued, it was cruel not to allow them to use it. Randall compared women being denied the opportunity to exercise their intellect to the fate of Tantalus, the character from Greek mythology who had been condemned to stand hungry underneath a tree laden with fruit that would remain forever out of his reach. Woman, wrote Randall, was "like Tantalus, placed in a situation where the intellectual blessing she sighs for is within her view; but she is not permitted to attain it."[1]

Randall was not alone in employing Enlightenment values to promote the expansion of educational opportunities for women in the new republic. Other advocates for female education used a variety of Enlightenment ideas to bolster their position. Some argued that a scientifically educated population was a boon to civilization, others that women possessed high intellectual capacities, and still others that knowledge was a source of personal satisfaction that should not be denied to women. Proponents of higher education for women also used the language of the Enlightenment-derived ideology of civic republicanism, which posited that for the new republic to survive, all citizens must be educated in reason and virtue. These arguments for further education in the early republic, including those that linked education to social progress, were the same for both women and men.

Other arguments regarding higher education were specific to females, including the basic question of whether women were capable of advanced education. One set of gender-specific arguments centered on women's relationships and roles in families and society. Prescriptive literature portrayed women as possessing unique powers to shape the behavior of men, and therefore assigned women an important societal role in cultivating civic virtue. Advocates of higher education for women who subscribed to this view argued that the benefits would be manifold: the husbands of educated women would be more virtuous, their children would be educated to civic responsibility, and the society as a whole would be elevated. In addition to broad social relationships, women's roles within marriages figured prominently in the discourses of education. The early national period was one in which ideas of marriage were undergoing change, and female education reflected diverse views. Those in favor of the new idea of companionate marriage viewed women's education as essential to a mutually fulfilling relationship. Those who held more conservative views saw education as a threat to the institution of marriage, fearing that it might "un-sex" women and make them unfit for domestic work. People on either side or in the middle on the question of what marriage should be like still might agree that some form of education was necessary in order for women to be well prepared for motherhood.

Societal benefits and intimate relationships were only part of the story. Aside from all of these "social utility" arguments, another set of reasons was given for advanced education for women. The Enlightenment elevated the individual, so it is not surprising that women actively sought out higher education for reasons that had less to do with their relations to others, and everything to do with their own needs and interests. These women reached for the intellectual blessings for which Anne Randall sighed.

All of these views flourished in essays, articles, and speeches published in newspapers and magazines in the immediate post-Revolutionary years— during which both Enlightenment and republican ideology held strong currency in the United States—through the first decade of the nineteenth century.[2] After 1815, fears both of internal rebellion and of foreign attack diminished, and the nation turned its energies and attention to issues of increased industrialization and urbanization, territorial expansion, religious revival, and programs of social reform.[3] These diverse factors contributed to the continued growth of women's education in the antebellum era, which I discuss in chapter 4.

In the early national era, much of the discourse on education for both men and women centered around the theme of civilization. Periodical literature and essays of the early national period repeatedly emphasized the theme that education was a "mark of the progress of society."[4] "According to the neglect or cultivation of the sciences," wrote an anonymous author of an article in the *New Jersey Magazine* in 1787, "whole nations rise or fall."[5] Those who held this view did not believe that the acquisition of knowledge was important only for men; they deemed women's education just as significant to the rise and progress of civilization. In fact, some believed that the status of women, indicated in part by their educational attainment, was a measure of a society's civilization. This "four-stage theory" held that societies progressed from hunter to shepherd to farmer to merchant, and that the fourth, or mercantile, stage represented the pinnacle of development. In the primary, or hunter, stage, the savagery of the society was evidenced in part by the low status of women, who were essentially "the servants or slaves of the men."[6] In a mercantile society, at the other end of the continuum, the advancement of civilization was seen by "the gradual advance of the female sex to an equality with the male sex."[7] That equality was limited: it was a social equality, not a political, legal, or occupational one.[8] Those who subscribed to this model were not arguing for women's right to vote, to hold political office, or to enter into the professions. They were, instead, making an argument about the value a society placed on women. Adherents of this belief about the stages of civilization supported advanced education for women because they viewed educational attainment as a major indicator of social equality, and social equality between the sexes as an indicator of civilization.

Supporters of female education also used arguments based on a belief that the inculcation of scientific, rational thought counteracted superstition and bigotry and promoted true Christianity.[9] Education, according to this argument, would pull religion "from the gloomy reign of Paganism and superstition" to full splendor.[10] For these reasons, and in light of women's crucial role in the maintenance and dissemination of religion, educators argued that science was as important for women as for men. One example of the association of women with science is the seal of Poor's Young Ladies' Academy in Philadelphia, which depicted a stack of books partially encircled by a line on which was written "the Path of Science." The Path led to a picture of the "Eye of Science emitting its rays over the whole."[11]

But were women capable of scientific study? European scientists and political theorists had been debating the question of female intellectual capacity for years. It was obvious that women as a group did not have the roster of learned accomplishments that men did. Was this because of mental inferiority, or lack of opportunity? Various writers answered the question differently, reflecting a range of views. Some believed in intellectual equality, while others held a belief that women were capable of some learning but could never reach the levels attained by men. Some writers argued that the minds of women were different from men's—not necessarily inferior, but different, nonetheless, with different strengths, weaknesses, and capacities. Still others held that women neither could nor would want to learn much beyond that which was absolutely necessary to run their homes efficiently.

Advocates of women's high capacity for learning could find support for their views from several prominent Enlightenment philosophers. John Locke, whose theory of child development had a great impact in the United States, believed that males and females had equal potential. Published debates about women's intellectual abilities frequently drew upon Locke's ideas regarding environmental influences.[12] Locke believed that an infant's mind was a *tabula rasa*, or a blank slate. Parents and teachers, then, had tremendous influence over a child's mental development. If women did not appear to be intellectually strong, it was because they had not been trained in the use of their intellect. Locke therefore advocated the same education for girls and boys, since both were equally capable of using the power of reason.[13]

Two other Enlightenment philosophers also supported beliefs in intellectual equality. René Descartes, the French philosopher and mathematician, believed that proper training could eliminate any differences in individual intellectual capacity. He did not believe that women had inferior reasoning powers, nor did he believe that women and men had distinctive mental or moral faculties. François Poullain de la Barre, an ex-Jesuit scientist, proclaimed that the mind had no sex. According to him, the study of anatomy proved that the only differences between men and women were their reproductive organs; their brains, he contended, were exactly the same. His book, originally published in France, was first published in English in 1677 under the title, *The Woman as Good as the Man*, and reprinted in 1751 in the book, *Beauty's Triumph, or the Superiority of the Fair Sex Invincibly Proved.*[14]

Those who believed in intellectual differences between the sexes could find support from European philosophers too. Jean Jacques Rousseau believed that the effects of reproductive organs pervaded every aspect of one's constitution and, therefore, the brain, along with everything else, was sexed. Based on this assertion, followers of Rousseau developed the concept of sexual complementarity, which contended that women and men were fundamentally different, and that they could achieve equality only in the sense of perfecting the performance of their distinct roles.[15] Some people believed that male and female minds were composed of different essential substances, resulting in different intellectual capacities. "The female mind," said the author of a 1793 essay in the *Massachusetts Magazine*, is "composed of more feeble and delicate materials."[16] The theory of complementarity had many proponents, and comparisons were made frequently: men think profoundly, women feel profoundly; the male mind is penetrating and contemplative, the female mind is lively and rapid; man surveys and observes, woman glances.[17] William Alexander, author of a history on women, pronounced in 1796 that nature simply did not intend women for intense and severe studies.[18] Some women agreed that females had a comparatively meager intellectual capability. Religious writer Hannah More urged women to relinquish their "pretension to strength of intellect" and to admit that the female mind is not as capable of science as the male mind. Yet More also called attention to the unpopularity of this belief, saying ruefully that she "fears it will be hazarding a bold remark, in the opinion of many ladies" to argue her point.[19] Her defensive tone indicates the extent to which other women were asserting their intellectual abilities, if not their intellectual equality with men.

As the discussion progressed in the early republic, some people insisted that intellectual capacity was not the issue at all. One critic wrote that many women "with strong mental powers, are little inclined to the trouble of exerting them. They love to indulge a supine vacuity of thought . . . because it requires no effort."[20] A commentator in the Maryland journal *The Key* chastised women for their obsession with fashion, saying that "all the labored efforts" of various journals that tried to uplift women could not "pluck one feather from a ladies [*sic*] head dress, or add one sentiment of virtue to her mind; . . . [she] remained as before, the giddy and gaudy object of her unbridled pursuits."[21] To these critics, women might be capable of acquiring knowledge but were too swept up in superficiality to bother using their brains.

Others countered these arguments by asserting that it was nurture, not nature, that kept women from fully developing their minds. The author of a 1792 article in *Lady's Magazine* acknowledged that men might perhaps be of more sound judgment than women, but contended that if so, it was due to the advantages of education and travel. This discrepancy was patently obvious, she stated, from the observation that until the age of 13 or 14, girls were "every where superior to boys." After that point, though, boys reaped the benefits of advanced education, until the age of 23 or 24, whereas girls stopped their education, "such as it is," at the age of 18. "He has all the fountains of knowledge opened to him. . . . His talents are always on the stretch," while girls are allowed nothing more than rudimentary education. Effectively making a case against the existence of innate differences, the author proposed finding a boy and a girl, neither of whom could read and neither of whom had been out of their own village. "I question very much," the author concluded, "if his discretive faculties will be found to be stronger than her's [*sic*]."[22] Writers frequently repeated this proposition, which assumed no sex-based difference in mental faculties and stressed that observed differences were rather due to variations in opportunity and expectations.

Other men and women echoed the Lockean position that with equal opportunities, women would prove themselves equally capable. The principal of a female academy in Philadelphia asserted that women were "fully capable of profound depths in every science," if only they had not been excluded from opportunities to develop such depth.[23] Excerpts from Count d'Hartig's *Miscellanies in Prose and Verse*, printed in the *New York Magazine* in 1790, expressed the sentiment that it "is a prejudice to imagine, that even the most abstruse sciences are incompatible with the genius of women." Like many articles and speeches from this era, the article listed examples of educated women throughout history.[24] One author expressed the hope that women would take up the study of science to the extent that "the toilet and looking glass shall be neglected for the acquirement of science."[25] Science, this author implied, was well within the grasp of most women and, moreover, as appropriate for a woman to use as a mirror. Elizabeth Hamilton, a famous writer of the period, noted that "Nature . . . has endowed us" with intellectual faculties, so women should exercise them. "Why these should be given to us as a *sealed book* which ought not to be opened, I confess I cannot comprehend," she wrote.[26]

A host of writers, from the famous Mary Wollstonecraft to the Philadelphia schoolgirl Priscilla Mason, argued on behalf of the belief in women's intellectual equality with men. Some of these writers primarily sought increased educational opportunities and respect for their intelligence without advancing a broader equal rights agenda. Abigail Adams was one who desired better education for women without also envisioning a more public role for them. In a private letter to her cousin John Thaxter, who was secretary to John Adams when Adams was on a diplomatic mission in France, Abigail wrote that she found the lack of female education "mortifying." Everything is done for the sons, she wrote, "whilst the daughters are totally neglected," even though females have a "common share of understanding" with men. She did not understand why men would "wish for such a disparity in those whom they one day intend for companions and associates."[27]

The Revolution itself stirred women's thinking on this issue. Many women became engaged in the political thinking of the Revolution, and sustained this interest after the war. Historian Cynthia Kierner provides the example of Elizabeth Steele, a North Carolina tavern keeper, "who read Tom Paine, followed the progress of troops and diplomacy, and lent gold and silver to the Continental Army"; women like Steele "were unlikely to lose their interest in political affairs" once the war ended. As they too sacrificed for the war, they wanted their thinking on such matters to be respected. Another Southern woman was irked by men who "treat women as Ideiots [*sic*]" instead of respecting their abilities to understand current events. South Carolinian Eliza Wilkinson, who suffered through the British occupation of Charleston, asserted that women's minds "can soar aloft," and that they "can form conceptions of things of higher nature; and have as just a sense of honor, glory, and great actions" as the men who would suppress women's "liberty of thought."[28]

Others championed increased educational opportunities for women along with expanded political and economic rights. One of these was the popular New England poet, essayist, and playwright Judith Sargent Murray, who asserted that the "*idea of the incapability* of women, is . . . totally *inadmissible*" [*sic*]. Once women's capability was admitted, she argued, women would have "effectually establish[ed] the female right to that *equality with their brethren, which, it is conceived, is assigned them in the Order of Nature*."[29] Several female academy students harshly critiqued a system that kept women subordinate. Priscilla Mason, a student at Poor's Young Ladies' Academy of Philadelphia, gave a salutatory oration in 1793 in which she said,

> Our high and mighty Lords (thanks to their arbitrary constitutions) have denied us the means of knowledge, and then reproached us for the want of it. . . . They doom'd the sex to servile or frivolous employments, on purpose to degrade their minds, that they themselves might hold unrivall'd, the power and pre-eminence they had usurped.

Mason happily noted that educational doors were beginning to open, and she named a set of women who had demonstrated their abilities in the field of science. Opening the doors to advanced education was not enough, however, because even if women became proficient in mental faculties, "where shall we find a theatre for the display of them? The Church, the Bar, and the Senate are shut against us. Who shut them? *Man*; despotic man."[30]

Some framed their demands for increased access to higher education in revolutionary terms: those in opposition were tyrants, denying women their human rights.[31] One young woman who objected to a lack of education directly compared the denial of educational opportunities for women to the British tyranny over the colonists. In 1794, the *New York Magazine* printed "On Female Education," written by a student at a female seminary in New York. The writer opened the essay by exclaiming that she was "struck with amazement to observe the material difference in the education of the sexes."

Asserting as common knowledge that men and women have the same natural abilities, she was unable to understand why equal opportunities were not forthcoming. She answered her own question by saying that with education, women would "establish our rights, and trample on the despised flattery of those who wish to keep us in the base chains of ignorance." She hoped that Americans' recent battle for liberty would keep men "from exercising that despotism over us, which they so openly despised in their master," and that men would "wish to see the fair sex on an equal footing with themselves, enjoying all the blessings of freedom."[32]

How one answered the question of women's intellectual capabilities carried social ramifications. One essayist drew the lines clearly. If an author "should assert the mental inferiority of the female sex," he wrote, he "would be upbraided by the one party, as the advocate of tyranny, and the slave of prejudice; and on the other hand, . . . [he] who maintains the intellectual equality of the sexes, will hardly escape the opprobrium of a traitor to his party, who . . . fights the battles of the enemy."[33] Many women and men believed in "the intellectual equality of the sexes" and championed advanced education for women. Proponents of women's education applied elements of an Enlightenment heritage that placed a high value on education. Believing that women were as capable of intellectual attainment as men, they pushed for more opportunities for women to develop their intellects. Like the anonymous New York seminary student, some invoked the antityranni-cal spirit of the Revolution with all its social and political authority to denounce those who would exclude women from education and link those opponents to the vanquished and now despised Tories. Most people, though, agreed that both men and women should acquire some basic education, and many of the arguments for this position centered around themes of civic duty and the preservation of the republic.

# Creating a Virtuous Republic: The Social Utility of Female Education

Leaders of the new republic had many reasons for encouraging the education of both women and men. To some, the newly knit republic seemed con-stantly in danger of unraveling. For the first generation of its independence, according to historian Harry L. Watson, the United States was "obsessed with preserving its republican experiment from the danger of internal col-lapse."[34] Such an obsession was not groundless. For instance, in 1786 Daniel Shays led a rebellion of 2,000 Massachusetts farmers who were threatened with foreclosure of their mortgaged property; the insurrection briefly closed the courts and threatened a federal arsenal. The Whiskey Rebellion of 1792–1794, in which a group of western Pennsylvanians openly and violently opposed a tax on whiskey, was another frightening indication of social disorder.[35] One response to fears of unruliness and social chaos was that political and educational leaders focused on the necessity of inculcating virtue and reason in both the male and the female citizens of the new country.[36]

Although supporters of increased attention to formal schooling in the new republic differed among themselves on the extent to which education should be a national or state responsibility, they all held a strong belief in the power of education to create a moral, intelligent, and unified citizenry.[37] Noah Webster, an ardent nationalist famous for his spellers, readers, and dictionary, hoped that America would separate itself from European culture and literature and create its own unique culture and language.[38] A system of public schools, he advised, would forge a distinct American identity and inculcate in children "an inviolable attachment" to their country.[39] He objected to states providing colleges and academies for the sons of "people of property" without "instructing the poorer rank of people even in reading and writing." Webster argued that education should be adapted to the principles of government: in a republic, "every class of people should *know* and *love* the laws."[40] In addition to studying reading, writing, and arithmetic, he recommended that young people of both sexes be taught "submission to superiors and to laws, the moral or social duties, the history . . . of their own country, the principles of liberty and government."[41] Webster and other theorists urged that females as well as males participate in this inculcation of virtue.

Webster was one of many who believed that women wielded extraordinary power over men. According to this view, women had a civilizing influence.[42] A commencement speaker at Columbia College said in 1795 that women had the power to mold men's taste, manners, and conduct, "change their tempers and dispositions," and inspire men to noble deeds. Desirous of winning female approbation, men would become whatever women wanted them to be. Women, "as patriots and philanthropists," could shape men's behavior "conducive to the glory of the country." The author of an advice article published in the *American Magazine* in 1788 instructed women about the nature of their power: "you polish our manners—correct our vices—and inspire our hearts with a love of virtue."[43] Webster averred that men "have been restrained from a vicious life" by their "attachment to ladies of virtue." Women's sway covered a broad scope, as they made their impress, not only on one or two individual men, but on the whole country. As an essayist for the magazine *Universal Asylum* put it, women had a powerful influence in "controlling the manners of a nation."[44] One justification for female education, therefore, was that it would help women use their power for good.

This emphasis on the "civilizing" role of women implied a greater equality in marriage, one of the primary sites in which women's influence was enacted. Relationships between men and women within the context of marriage were very much at issue in the post-Revolutionary years. Earlier in the eighteenth century, cultural ideals of marriage shifted away from the patriarchal model in which an authoritarian father in his home represented God's dominion over his kingdom. Young people increasingly chose their own marriage partners rather than simply obeying their fathers' commands in this regard. Further, as the eighteenth century progressed, the basis for choosing a mate leaned more and more toward love and affection.[45] Even in the South, where patriarchal attitudes continued to prevail in the plantation system, companionate marriages were the ideal for many young couples.[46] The Revolution only heightened these tendencies. Americans had overthrown the yoke of a monarch; the old metaphor of absolute authority resting in the hands of a paternal figure no longer held political resonance. The new republican model needed a new metaphor, and the one that many writers adopted was that of a consensual, affectionate marriage, "freely entered into, without tyrannical interference."[47] The author of an article in the *New York Magazine* asserted that the "idea of power on either side, should be totally banished," and that marriages founded on affection were the happiest.[48] In this view, husband and wife should have a mutual regard, and should be thought of as matrimonial partners with different, but equally valued, strengths, attributes, and duties.[49]

An article in Philadelphia's *Lady's Magazine* in 1792 demonstrates the belief in marriage as a mutual partnership. In this article, the author urged that the word "obey" be taken out of the marriage service. Identifying herself only as a "Matrimonial Republican," she objected to the custom of women promising to be obedient to husbands, arguing that such an unqualified vow bound women as slaves. "There is no woman," she stated emphatically, "born to be a slave." Misery in marriage, she argued, stemmed from

> the assumption on one side or other of absolute power. Marriage ought never to be considered as a contract between a superior and an inferior, but a reciprocal union of interest, an implied partnership of interests, where all differences are accommodated by conference.[50]

This author, then, associated republicanism with equal partnerships within marriage.

Men, some writers contended, would be happier with educated wives. The author of a 1797 article lauded men's choice of educated women for wives, describing

> the pleasures which men of genius and literature enjoy in a union with women who can sympathize in all their thoughts and feelings; who can converse with them as equals, live with them as friends; who can assist them in the important and delightful duty of educating their children; who can make their family their most agreeable society, and their home the attractive centre of happiness.[51]

Educated women could enrich the lives of busy men whose "labour [sic] and fatigue," as John Swanwick said, gave them little time for "cultivating the finer and more delicate branches of education." Swanwick, a prominent Philadelphia banker and trustee of Brown's Young Ladies' Academy, rhapsodized about the attainments of the students, adding his view of theology and marriage. Men, he said, look forward to "the acquisition" they would make when married to such women. "To give us happiness . . . was the intention of a benevolent Deity, in adding women to the society of men. You, I am sure, will make it your study" to carry out the divine design.[52] Women who filled their minds with knowledge and who were trained with the discipline of study had, according to the *Lady's Magazine*, a "charm that outlives beauty." Gone were the days when men shared the "silly prejudices of our forefathers who, in their wives, looked for upper servants, but seldom rational companions."[53] A South Carolinian wrote that "a man who looks upon a wife as but an upper servant" is not deserving of the "company and conversation of a virtuous and sensible woman."[54]

Two women writers of the 1790s cautioned men that it was a grave error not to value education in a woman. One wrote in the *Universal Asylum* that men "contribute to their own wretchedness when they neglect the culture of our minds." Mental qualities, she said, are the source of life's truest enjoyments, and "men are seldom brutish to such a degree, as long to enjoy the company of women who can only gratify the lowest appetite."[55] Another woman used the most flagrant flattery to convince men of the benefits they would accrue when they valued education in a woman. "By degrading our understandings," she wrote, "you incapacitate us for knowing your value . . . how impolitic to throw a veil over our eyes, that we may not distinguish the radiance that surrounds you!"[56]

The ideal of mutuality in marriage, in turn, implied a certain support for female education. In order to help create and sustain the stability of the new republic, as well as to be equal partners and ideal companions, women and men needed similar education. Reflecting this belief, several educators published model curricula. In Philadelphia's *Weekly Magazine*, John Hopkins spelled out a plan for female education that bore remarkable similarity to academy education for men. His 1798 plan included writing, arithmetic, grammar, reading, rhetoric, composition, geography, the use of globes, ancient and modern history, natural history, natural philosophy, logic (or what this author referred to as "The Art of Thinking"), and moral philosophy.[57] Erasmus Darwin, the British philosopher, botanist, and grandfather of the famous Charles Darwin, published a plan for female education in which he recommended the following subjects: writing, reading, grammar, languages, arithmetic, geography, history, natural history, literature, and mythology.[58]

Not everyone favored a more companionate model for marriage, however. Some reacted against this view and instead asserted a husband's claim to dominance in a household. As Mary Beth Norton remarks, those who advocated the ideal of mutuality in marriage coexisted with those who promoted female subordination within marriage.[59] Those who held an ideal of male dominance within marriage were more likely to favor female education that was limited to the bare essentials of basic literacy, domestic skills, and perhaps the female "accomplishments" of music and painting, depending on the socioeconomic class of the people involved. More education than that, according to some, rendered a woman decidedly unfeminine. A woman who has too strong "a predilection for the sciences," said a writer for the *American Magazine* in 1788, has "quit her own department," thereby offending her husband if she has one, or deterring any potential suitors.[60] Several writers connected education with symbols of masculinity, and asserted that neither education nor any masculine symbols belonged anywhere near women. One man wrote that "we should perhaps grudge [women] the laurels of [literary] fame, as much as we do the breeches."[61] A writer for the *New York Magazine* stated, "A woman with a beard is not so disgusting as a woman who acts the free-thinker."[62] Keep classical education for men, warned an essayist for *Lady's Magazine*, else we might see battles "fought with superior skill" by women, as "the Ladies make rapid advances towards *manhood*."[63]

With revolutionary rhetoric flying on both sides of the Atlantic, scores of women vociferously objected to a social and political system that denied them some semblance of equality within both marriage and society. Demonstrating the cultural value of education, at least among the elite, one of the most frequent complaints women lodged against men was that they arrogated knowledge to their own domain. Some women minced few words about what they considered to be the real reasons men did not prefer educated women. Anne Randall wrote, "A thinking woman does not entertain him; a learned woman does not flatter his self-love, by confessing inferiority; and a woman of real genius, eclipses him by her brilliancy." She suggested that men worried that if they allowed women "to participate in the intellectual rights and privileges" that men enjoyed, there would be no one left to "arrange our domestic drudgery."[64] Or, as one woman angrily put it in an article in the *New York Magazine*, men feared that "by enlarging and ennobling our minds, we shall be undomesticated and unfitted, for mere household drudges."[65] In spite of these vituperative skirmishes, most writers on female education supported the view that virtuous women helped create virtuous men, and therefore a healthy republic.[66] Most further agreed that advanced education would help prepare women to shoulder this responsibility, and would make women better companions for men.

Preparation for good motherhood also figured in the arguments of proponents of female education, although historians have probably exaggerated the importance of this issue.[67] According to some theorists, women wielded considerable power as the primary rearers of children. When Webster discussed female education, he asserted that it should help women teach "virtue, propriety, and dignity" to their children.[68] Benjamin Rush, Revolutionary Army surgeon-general, signer of the Declaration of Independence, and advocate for female education, expressed the idea that women should be educated in order "to concur in instructing their sons in the principles of liberty and government."[69] Yet close analysis of Rush's essays reveals that he advocated female education to prepare women for multiple duties and responsibilities. Rush did not see motherhood as women's only, or even their primary, role. He expected women to put their education to use in helping to manage their husbands' businesses, run efficient homes, contribute thoughtful conversation to social groups, and improve their own health and happiness. Rush's Enlightenment beliefs in science drumming out superstition, as well as his concurrence in prevailing ideas of women's relationship to men, were at the forefront.[70] Nor was Rush the only writer of the era to emphasize nonmaternal female roles. Historians have taken note of this in the past decade, and have challenged the dominance in the historiography of the construct of "republican motherhood." As Ruth H. Bloch contends, "there was far more literature on courtship, marriage, and the social utility of female education than on motherhood."[71] Social utility arguments, then, suggested that education was necessary to prepare women for a range of duties as wives, mothers, and citizens. Other arguments focused on the personal satisfaction that would stem from education.

## Seeking Great Blessings: Personal Rewards of Education

The articles and essays recommending the "social utility" of women's education have received much attention from scholars, whereas the equally important theme of education for women's self-interest and fulfillment has been neglected. A full consideration of female education that gives due weight to the idea that women had intellectual interests that deserved realization radically alters the picture of this period. The social utility arguments extended women's opportunities but also maintained women in an inferior position by emphasizing the service that educated women could provide to others. In contrast, arguments from self-interest assumed that women, like men, were fully human in the Enlightenment sense—that they had a capacity for intellectual achievement, that that capacity was what distinguished them from the animal world, and that their intellects were equally worthy of cultivation. Self-interest as a reason for female education had implications that went well beyond the curriculum.

Proponents of female education put forth justifications beyond increasing women's potential value to men and the country. They offered other reasons, reasons that were more directly tied to women's own self-interest. One prime consideration was the nongendered nature of the pleasure of learning. Closely associated with this was the belief that education would improve one's relationship with the divine, and this reason, too, applied equally to men and women. A third, and extremely gender-specific, argument for female education related to the supposition that marriage was likely to be more difficult for women than for men, and education would help women survive trying circumstances. Finally, advocates of female education touted the practical benefits that would accrue, such as improved household management and the potential for self-sufficiency. In fact, the model republican woman was not one who was passive at all; instead, she was self-reliant, competent, and confident in her abilities.[72]

Writers often extolled the pleasures of learning, for both women and men, and linked learning with happiness.[73] "Education is the groundwork on which the Temple of Happiness may rise," said one orator at a female academy in 1787.[74] In an essay published in 1794, the anonymous New York student gushed,

> Oh! learning, thou art one of the great blessings mankind can enjoy. . . . Neither can I conceive that learning was intended merely for the improvement of one sex . . . for those who have been the most learned, have been the most happy.[75]

A writer for the *New York Magazine* advised women in 1790, "If happiness be dear to you, attend to the cultivation of letters."[76] That same magazine published in 1795 an article on the study of arts and sciences in which the author contended that "the desire of knowledge is planted in every human breast; it is as natural to us as reason," and from the study of science "we derive the principal delights of life."[77] Hannah Adams, who wrote a history of New England that she also abridged for use in schools, shared this view. "My first idea of the happiness of heaven," Adams wrote in her memoir, was of "a place where we should find our thirst for knowledge fully gratified."[78]

These sentiments were expressed privately, as well. Jonathan Steele, a North Carolinian who served as comptroller to the U.S. Treasury, wrote to his daughter in 1800, "Go on my dear child, to improve your mind—knowledge is the best resource at that time of life when youthful pleasures fail. It makes man, or woman, in a certain degree independent of circumstances, and affords them consolations where the rational only can seek it with success."[79] Similarly, Joyce Myers wrote to her stepdaughters, who later would become well-known teachers in Warrenton, North Carolina, praising them for their progress in school. She urged them to continue their studies, saying, "the pleasure resulting to yourselves and the advantages attendant on Education will no doubt all conspire to impell [sic] you with ardor steadily to pursue this pleasing avocation."[80]

Women in the early republic pursued learning with great ardor, whether or not formal education was available. Hundreds of women organized reading circles and literary societies in which they read and discussed everything from theology to history to astronomy. In Boston's Gleaning Circle, founded in 1805, the members took turns presenting formal essays to the group, drawing from what they had read and applying those lessons to the issues of the day. Scores of other reading circles also followed the same pattern. According to historian Mary Kelley, in virtually every town or village across the country, women gathered to read and write together.[81]

The belief that learning would bring women and men a richer understanding of God was also widely held. Studying the natural sciences provided opportunities to "admire the works of God"; through detailed observation of works of nature, students saw "a thousand instances of the perfection of the Deity." Such an improved understanding rendered students "better qualified to obtain the favor of God," and increased the likelihood of their promoting virtue. Education also increased their potential for heavenly bliss because "the more exalted and refined the mind is in this world the more perfect and sublime will be its happiness in Heaven."[82]

While some proponents of female education expressed those lofty hopes that education would lead women to God, others more humbly suggested that it might, at the least, prevent dissipation, vice, and despondency. Many writers depicted women's lot in life as lonely, tedious, and filled with distressing circumstances. Education, some thought, would help ameliorate such unpleasantries. It would "often enable them to avoid, and always to bear, the inconveniences of domestic life," and "remove the necessity of resorting to trifling, perhaps criminal amusements" that women might easily fall prey to.[83] Ann Negus, a Philadelphia female academy student, wrote in 1794 that men have more opportunities to "extricate themselves from misfortune" than do women, who are more prone to slip into dissipation, or give in to "whims of fancy." This problem, she felt, "could be resolved by regular and classical education."[84] Another student that same year maintained that education was more necessary for women than men, because women "lead a more solitary life, and must, unavoidably, sometimes fall into melancholy and dejection if not supported by a good education, which would enable us to pass those pensive hours in contemplation and writing, which would . . . sweeten adversity, and soften the cares of life."[85]

Although marriage might add greatly to men's happiness, some women expressed the belief that it was not always such a good bargain for females. Ann Negus, the student mentioned earlier, reiterated the pitiable condition of women in a 1794 address. Negus averred that the sufferings of men in battle were nothing compared to the sufferings of women in unhappy marriages. Soldiers, at least, had each others' company and an assortment of amusements, received the applause of their fellow citizens, and had the satisfaction of looking back with pride on their service. Women, however, had none of these rewards to mitigate their sufferings, and nothing "so soon sinks the mind into hopeless despondence, as contemptuous neglect."[86] An anonymous writer for *Lady's Magazine* declared that marriage was nothing more than a lottery, and many of the husbands given as prizes were simply "John trots, fond fools, drunken fools, unfaithful fools, . . . stupid fools, rich fools, . . . old fools, young fools, handsome fools, ugly fools. . . ."[87] These negative assessments of marriage contributed to the contention that, while education obviously would not ameliorate all the difficulties seemingly inherent in married life, it would make life more palatable. Enriching women's minds and hearts would provide a source of solace during dismal days.

In addition to such intangible benefits, many writers also saw practical purposes for women's education. Rather than rendering women unfit for domestic drudgery, as some feared, proponents argued that education would help women fulfill their domestic obligations "with grace and dignity." Lack of education, they argued, not too much education, would interfere with the performance of women's work. Those whose studies "have been confined to Cookery," argued one writer, fight with the servants, "have ruined their health, spoiled their tempers, neglected their persons, laid waste their minds . . . and muddled away their time and money in disorderly management."[88] In an argument that would be often repeated in the early nineteenth century, proponents of advanced education insisted that educated women made the best daughters, wives, mothers, and citizens.

Within routine household management were many tasks that required literacy and numeracy. Women needed to be able to buy and sell advantageously, keep accounts, compose letters, and manage estates in the event of their husbands' or fathers' absence through travel, illness, or death.[89] A man's solvency in many ways depended on his wife's regulation of the household economy, especially when she often served "as treasurer of his family." Not only did they need the technical skills of writing and book-keeping, but they also could benefit from the mental discipline of other subjects. Studying logic and philosophy, for instance, would teach women to think rationally, which, in turn, would aid them in all their tasks.[90]

Education had an important purpose for women beyond improving the dispatch of their domestic duties. Education, some argued, could also lead to economic self sufficiency for widows or never-married women. Judith Sargent Murray essayed the hope that the term "helpless widow" might be rendered inapplicable, and that a recently bereaved woman might be able to "*assist* as well as *weep* over her offspring." Women, she argued, ought to be taught to "depend on their own efforts" rather than rely on their husbands to support them. Marriages, she and others believed, had a greater chance of success if they were formed for other than pecuniary reasons.[91]

An even more compelling reason for women to be self-supporting was their desire for self-determination. Murray argued that "the freeborn mind . . . naturally revolts" against dependence, and that this was no less true of women than men. "Aim at making yourselves so far acquainted with some particular branch of business," she told women, that you need not sink into dependence.[92] An anonymous writer in the *Monthly Magazine*, in an essay that was reprinted in the *New York Magazine* in 1797, echoed Murray's view. Eschewing the "enfeebling effects" of a system in which women "are uniformly educated" for dependence, this writer promoted the possibility of education for a self-sustaining vocation. Female education, according to this essayist, should include the attainment of a useful trade by which women might "gain an honest and honorable independence, and be freed from the disgraceful necessity of bartering her person to procure a mainte-nance."[93] Another anonymous writer whose essay was published in the *American Museum* complained that, lacking instruction in any "art by which they can obtain subsistence," women, especially widows, "are too generally reduced by the last degree of wretchedness and misery, to abject poverty."[94] Yet another essayist recommended the Dutch system of education in which both girls and boys were "initiated in some line of industrious avocation . . . [to] render them virtuous and independent citizens."[95] Such professions might involve knowledge of arithmetic and bookkeeping.

More women would be likely to need such a trade if they followed the advice literature that cautioned them not to marry hastily, instructing women that no marriage was better than a bad marriage. An anonymous writer in 1792 asserted that many unmarried women "act very prudently in declining entering matrimony with such suitors as they often have."[96] One advice book recommended telling girls about the examples of respectable and "prudent" women who "pass an easy, independent" single life, "as happily as any widow, and assuredly more so than many a married woman."[97] A Massachusetts poet wrote an ode to singleness that was based on the metaphor of herself as a republic that, like the new country, desired only independence; the pledge of allegiance she made was to the "innocent freedom that dwells in my mind," rather than the "fetters [that] bind" in wedlock.[98] More women would enter the "cult of single blessedness" in the antebellum era, and educational institutions played a part in their ability to achieve self-sufficiency.[99]

The discourse on advanced education for women in the early republic, then, reflected tensions between the rhetoric of human rights, revolutionary ideologies, and Enlightenment ideas about intellectual equality, on the one hand, and fears of social instability brought about by individuals demanding their rights, on the other hand. Filled with revolutionary rhetoric, some women and men quickly saw an analogy between colonists fighting for independence from Britain and women fighting for their own set of rights. Women's lower status was painfully apparent to some, and journals and books carried the theme of the insult of women's subordination. Some argued for female equality, while more argued for a modest expansion of women's rights, including the right to education.

One set of justifications for advanced education focused on women's relationships. These arguments centered around the positive effect that educated women would have on men, their families, and on the society as a whole. Enlightened women would inspire men to better behavior, morals, and industry, and would create homes that would be pleasurable for men. Educated women would be better wives and mothers. Another set of arguments revolved not around ways that women could put their education to use in service to men and children, but around ways that education could serve women themselves. Women, especially, frequently articulated the delight that education would give them. Intellectual pastimes would not only bring women personal satisfaction, but they also would keep women from slipping into the despondency to which their lives might otherwise lead. Finally, both women and men cited multiple practical reasons for women's education, not the least of which was creating the potential for financial independence.

Anne Randall's "plain and rational question" was answered largely in the affirmative: a woman was a human being, and was entitled to the rewards of an education. But exactly what sort of education did women receive in this era? To better understand women's experiences, one must first enter the institutions of higher education and examine the curriculum and pedagogy of the academies.

# Chapter 3

# "Cultivating the Powers of *Human Beings*"

## Curriculum and Pedagogy in Schools and Academies in the New Republic

Elizabeth Hamilton, author of the popular *Letters on Education* (1801), was a strong advocate of advanced education for females. When someone suggested that a "triumph of reason over the passions" might be unattractive in a woman, she retorted, "I beg your pardon; I thought we were speaking of the best method of cultivating the powers of *human beings*. . . . In this I can make no distinction of sex."[1] Most writers on education in the early republic agreed with Hamilton. The majority of educators believed that both males and females were rational human beings who needed to acquire mental discipline and who were interested in learning about the world around them. Both the curricula and the pedagogical methods proposed by educational theorists in the new republic reflected these beliefs.

The academy was the chief institutionalized form of advanced education in this era, and it defies simple descriptions. For men, academies occupied an ambiguous place between grammar schools and colleges. For women, academies held a place quite distinct from colleges: academies were open to women whereas colleges were not. Beyond that, however, distinctions between academies and colleges were not clear. Uncovering actual curricula, as opposed to proposals for courses of study, is also complicated. Advocates for female education, as well as other educational reformers, published many recommendations for the adoption of ideal courses of study for both women and men. What actually was taught varied considerably.

Precise comparisons between men's and women's education are difficult to make because there was wide variation among institutions, and also because our knowledge of actual classroom experience is partial. There were differences, to be sure; boys were more likely than girls to learn Latin, or to be offered training in surveying and navigation. Girls' instruction frequently included needlework, which was never offered to boys. However, there was not a single, clear "male" curriculum or a single, clear "female" curriculum. Not all boys in these institutions studied the classics, while many girls did. Some academies for boys emphasized preparation for college, some emphasized training for a specific trade, and others offered a general education that included mathematics, science, and history. Similarly, some academies offered girls courses in classics, some offered training for trades (especially those of bookkeeper or seamstress), and many offered a general education that included mathematics, science, and history.

Two other aspects of education also reveal intriguing similarities. Women's education frequently has been portrayed as largely "ornamental," that is, confined to music, needlework, or the fine arts. However, clear distinctions between "useful" and "ornamental" subjects are virtually impossible to draw; "ornamental" subjects were offered to both women and men and were central to what teachers routinely taught in the period. Finally, pedagogical practices illuminate much about a prevailing belief in women's intellectual capabilities in the early republic.

# Academies

The first schools for females that taught subjects beyond the rudimentary ones of basic reading, writing, and arithmetic were "adventure" (or "venture") schools that sprang up in virtually every geographic region along the eastern seaboard in the mid-eighteenth century. These were small, private, and often temporary establishments, usually with a single proprietor or a husband-and-wife team who taught one or more subjects. Some schools accepted only female students, others only male, and still others accepted both. Instructors in adventure schools taught music and dancing; drawing and painting; French, Latin, and Greek; fencing; fancy needlework; and just about any other subject for which there might be a market. Some schools provided instruction in English grammar, writing, arithmetic, and literature. These schools, which appeared in larger towns before the Revolution and spread to smaller ones after the Revolution, were the basis for the later female academies.[2]

In the late eighteenth century, adventure schools faded away in favor of academies. While adventure schools were temporary establishments, opened in someone's home for as long as either teacher or students desired and just as quickly closed again, academies aspired to more permanence. In addition to their less ephemeral nature, academies offered a wider range of subjects. Under the venture school system, a teacher offered instruction in one or more subjects. Venture school students might take dance from one teacher and go elsewhere for French lessons, and perhaps hire yet another instructor for music lessons at home. Academy founders, on the other hand, intended to provide a complete range of courses under one roof (although, as we will see, there was no agreement on what constituted a "complete" education).[3]

The academy movement represented a shift in thinking about education. In the early eighteenth century, wealthy families hired tutors for their children, preferring home schooling to education in an institutional setting. This education generally included rudimentary subjects for both girls and boys, as well as college preparatory studies for boys. Those who followed this practice found support for it in the opinions of philosophers like the highly influential John Locke, especially in his view of a young child's mind as a *tabula rasa*. Locke believed that the purpose of education was primarily moral and religious, and that parents were responsible for inculcating morality in children. By sending away children to school, parents could not perform this role. He also argued that young people gathered in groups were bound to exert a bad influence on each other. "Vertue [*sic*] is harder to be got" than book knowledge, Locke wrote, "and if lost in a Young Man is seldom recovered."[4] For all these reasons, wealthy parents were more likely to bring tutors to their homes than to send their children to an institution in which someone other than the parents was in charge.

By the mid- and late eighteenth century, however, elite families began to abandon the custom of private home tutoring and chose instead to educate their children in academies. This change may have been prompted in part by an increased emphasis on science, the study of which required apparatus that few families owned. At the same time, educators began to express concern about young people whose parents indulged them far too much at home. Critics especially targeted mothers. A founder of the Academy of Philadelphia contended in 1750 that "Partiality and Indulgence render them [mothers] unfit to be trusted with the sole Care of the Children's Education."[5] Young people did not need to be coddled, he insisted. They needed to be challenged and corrected and only an impartial teacher could do this fairly and evenly.

After the Revolution, the broad and diffuse discourse of civic republicanism provided additional support for academies as opposed to home tutoring. One common theme in American writing was that collective life was necessary to form good republican citizens. Academies, according to their proponents, provided a mixing of social classes that would be favorable for the republic. Other observers found academies beneficial because of the possibility of forging friendships and connections, both business and marital, in formal school settings.[6] In addition, many citizens in the new republic saw the spirit of competition as a positive good, and one that could not be arrived at in the small and sheltered world of the family. As one proponent argued, people tutored at home "are apt to form too high an opinion of their own attainments or abilities." Observing firsthand the accomplishments of others would serve to diminish arrogance.[7] Educators harnessed this competitive urge in the pedagogical system of emulation, which later proved to be a controversial issue.

Many of the academies that flourished in the late eighteenth century received charters from state governments that lent an aura of permanence to the institutions, established them as promoting the public interest, and provided possibilities for state financial aid.[8] In Georgia, legislation in 1792 allowed each county to fund academies from the sale of confiscated property. North Carolina's legislature granted permission to incorporated schools to raise funds through lotteries in the late eighteenth century and through much of the nineteenth. New York established a Board of Regents to supervise all academies and colleges in the state; in 1792, the legislature granted the Regents a flat sum to distribute among the various schools as they saw fit.[9] With or without state or local funding, however, academies primarily depended on tuition to survive.

Academies rhetorically distinguished themselves from colleges by claiming to offer a more practical education. Throughout the colonial era, colleges routinely taught the classical curriculum of Latin and Greek, but critics of this system emerged by the mid-eighteenth century. Benjamin Franklin, for instance, advocated an English course for boys as an alternative to the classical collegiate course. His proposed course of study included English as well as modern languages (which would aid in trans-Atlantic business), along with the practical subjects of navigation, agriculture, and surveying. His ideas were only partially realized in the Philadelphia academy he helped found, which eventually became the University of Pennsylvania. Although they acquiesced to the popular demand for practical subjects, the trustees refused to abandon the classical course.[10]

Most men's colleges continued to teach classics in the late eighteenth century, but increasingly faced criticism for doing so. One line of criticism related to issues of efficacy and pedagogy. According to these critics, the value of classical literature was in the morals such literature taught. These morals, or civic virtues, could be taught as well, if not better, by being read in translation; there was no need for students to spend years studying to read these texts in their original languages. Besides, despite years of grinding application to the task, few students emerged with much facility with either Latin or Greek at all. The dreadful pedagogy, requiring endless memorization and recitation, simply did not work well for most students. By the late eighteenth and early nineteenth centuries, instruction in classical languages often was cited as an example of what was wrong with colleges.[11]

Another line of criticism contended that students should spend their time on more practical subjects. Noah Webster, for instance, thought that only those young men destined for the learned professions should spend much time learning "the dead languages." He especially deplored students spending more time with Latin and Greek than they did acquiring a thorough knowledge of English.[12] Webster's 1790 essay "On the Education of Youth in America" called for practical education, arguing that students should study those subjects that had direct relevance to their prospective lines of work. "Why," he asked, "should a merchant trouble himself with the rules of Greek and Roman syntax or a planter puzzle his head with conic sections?"[13] Leave the colleges to their classics, many seemed to be saying; the academies would teach students what they needed for real life.

Yet, despite the rhetoric to the contrary, the difference between colleges and academies was not always clear. Academies did in fact teach the classics, at the same time that colleges also broadened their curriculum. Academies had good reasons to teach classics. As historians Theodore Sizer and Robert Church point out, even if the classics were associated with elitism, it was an elite distinction that many young men wanted for themselves. Further, from the point of view of financially strapped academies, instruction in the classics was cheap compared to instruction in subjects that required apparatus for experiments and demonstrations. As a result, academies attended by men were as likely as colleges to teach Latin and Greek.[14]

Meanwhile, colleges attempted to move beyond the classical curriculum. Many post-Revolutionary college curricula included a foundation in the classics. The classics were augmented by studies in English and occasionally in modern languages. College curriculum included a heightened attention to mathematics and natural science, and, not surprisingly in the post-Revolutionary years, a higher regard for studies in history and government. Some colleges developed two tracks of study. The University of North Carolina, for example, by 1795 offered one degree program for those wishing a classical curriculum, and another degree for those wanting to concentrate on English and the sciences. As with academies, however, financial realities often impeded the implementation of these plans. Because colleges were small, the faculty might be composed of only one or two people. If those one or two were not learned in particular fields, those subjects would not be taught. Limited funds also meant limited scientific apparatus, which could be costly.[15]

Colleges and academies had much in common, given that they served similar functions to largely the same population, with one main distinction. Both offered education beyond the rudiments, primarily taught students who were between the ages of 14 and 25, and also accommodated students in the broad age range of 10–40. Both colleges and academies offered similar courses of study in spite of the general expectation that academies offered a practical education in contrast to the classical curriculum taught at colleges.[16] However, while no colleges admitted women until well into the nineteenth century, many academies in the late eighteenth and early nineteenth centuries were coeducational or female-only.[17] This difference may be explained by the ubiquity of academies, which were far more plentiful than colleges. In 1800, there were only 25 colleges in the United States, but there were hundreds of academies.[18] The increased number of academies meant more competition for students, requiring academies to appeal to a broader constituency. When colleges entered a period of great proliferation in the mid-nineteenth century, more colleges became coeducational.[19] During the period that colleges were attended by only a tiny proportion of the population, women were not admitted.

It is impossible to know how many academies were coeducational, or what proportion of the total they represented. Academies had much more local appeal than colleges, and so acceptance of women probably varied greatly according to locale. Clearly, coeducational academies abounded. One source documented 34 coeducational academies in New Hampshire, Massachusetts, and Maine between 1783 and 1805.[20] Connecticut had at least half a dozen coeducational academies during the same time, and many are known to have existed in the South, including at least five in North Carolina, and five in Richmond, Virginia.[21]

Whether the students were female or male, most academies in the 1780s and 1790s were day schools and not boarding schools. Students lived at home if they were able, or else lived with relatives or friends near the school. Educators and parents alike were leery of boarding schools. Teachers could exert tight control on students in the classroom, but the time spent out of the classroom was a different situation altogether. Benjamin Rush, like many other writers on education, was strongly against boarding schools, calling them "the gloomy remains of monkish ignorance." Young people needed to be separated from each other when not in school, thus securing "their morals from a principal source of corruption."[22]

The academy movement of the late eighteenth century cannot be defined easily, as there was tremendous variation from school to school. In some senses, academies were for an elite few. Attendance figures for this period are difficult to determine, but one estimate is that even by 1850 only a small minority—about six percent—of the population ever attended, let alone graduated from, an academy.[23] A family had to be wealthy enough to pay tuition as well as to forego a child's earnings during the years he or she might otherwise be employed. However, more young people in higher education attended academies than colleges. By that same date of 1850, academy attendance rates were as much as nine times that of college attendance rates.[24] Colleges prepared men for the "learned professions" of law, medicine, and politics. Academies prepared men for a much wider range of occupations and also provided women with opportunities for learning that otherwise remained closed to them for decades.

## Both Useful and Ornamental: Curricula in the Academies

Academies taught a wide range of subjects to young women and men, ranging from English to navigation to dance. Essays, addresses, and school circulars often tried to capture this breadth by assuring prospective students and their parents that the curriculum included both "useful and ornamental branches" of education. But what exactly did this phrase mean? Did "ornamental" education apply only to females (as some historians have assumed) and refer to drawing, painting, and music? Did the presence of these "ornamental" subjects suggest that women's education was highly elitist, and/or irredeemably debased academically?

Primary documents suggest that educators and other writers variously described a range of subjects as either useful or ornamental, and that there was no clear agreement on what subjects fell into each category.[25] The author of an article in the *Massachusetts Magazine* stated in 1791 that the "necessary" branches of education were reading and writing, and labeled the subjects of history, geography, and science as "ornamental."[26] Noah Webster, however, believed that arithmetic and geography were useful and necessary subjects for both men and women.[27] In 1789, a trustee of Poor's Young Ladies' Academy covered all bases and said that reading, writing, history, geography, astronomy, and arithmetic were "not only ornamental," but also "advantageous." Later that same year, another trustee of the same academy referred to the same list of subjects as being useful.[28] George Washington's stepson, John Parke Custis, referred to a classical education as "that easy that elegant that useful Knowledge which results from an enlarged observation of Men and Things."[29] What was ornamental to one was useful to another, and some saw utility in that which they also viewed as ornamental.

Even when writers agreed on which subjects were ornamental—and many concurred with the religious writer Hannah More that these included dance, music, drawing, and reciting—there was little agreement on whether the ornaments were salutary, and for whom.[30] Numerous journal articles associated such activities with aristocracy, and deemed them inappropriate in a republic.[31] Others thought, however, that these subjects were beneficial, and so advocated education in the fine arts for women. Maria Edgeworth, a popular author whose two-volume work *Practical Education* was first published in 1798, spoke of the great restraints under which women must live. Any attempt to further "diminish the number of [women's] employments, therefore, would be cruel." "Every sedentary occupation must be valuable," she wrote, "to those who are to lead sedentary lives."[32] Women, she felt, might find comfort and solace for their otherwise difficult lives through facility in one or more of the arts.

Some educational theorists also promoted ornamental education for males. Benjamin Franklin wished that boys could be taught "every Thing that is useful, and every Thing that is ornamental."[33] At the opening of Bristol Academy for boys in New Bedford in 1796, Simeon Doggett spoke of his hope that the students would learn "the ornaments of ease and grace of manner."[34] Both winners of the American Philosophical Society's 1797 prize for the best essay on education advocated teaching fine arts to men in colleges. Samuel Smith, urging "ornamental instruction" for men, argued that colleges should have room "for every species of genius. Every spark of mental energy should be cherished."[35] Samuel Knox's plan for a system of state colleges for men included instruction in "several ornamental accomplishments," such as dance and music.[36] Similarly, John Adams wrote to his wife Abigail, regarding their sons' education, that she should "make them disdain to be destitute of any ornamental Knowledge or Accomplishment." Abigail linked that which was useful to that which was ornamental when she admonished her son John Quincy Adams to "[i]mprove your understanding for acquiring usefull [*sic*] knowledge . . ., such as will render you an ornament to society."[37]

It is important, therefore, not to assume what was meant by "ornamental," as that term was used in various ways. Furthermore, "ornaments" were not necessarily gender-specific subjects in the late eighteenth century: the term was applied to aspects of both male and female education, and sometimes it was applied to subjects that we would term academic. The meaning of the term "ornamentals" remained ambiguous both as to content and as to gender. While some people scornfully regarded "ornaments" as frivolous, others regarded them as valuable, humanizing influences on men and women alike.

The primary sources of information regarding subjects taught in academies are newspaper advertisements and institutional circulars and catalogs.

Yet it is almost impossible to know how many students actually took any given subject. If the status of the school increased according to the higher level of subjects taught, the administrators may have listed courses that *might* be taught, in addition to courses that actually *were* taught, in order to put the school in a better public light. Since few academies in this period required students to take any specific subjects, the listing of a subject is no indication of how many students actually took it. This situation changed in the antebellum era, when some schools published the numbers of students, male and female, enrolled in specific classes. Subject listings also give no clue as to the depth or rigor with which the subjects were pursued. In spite of these caveats, newspaper and circular listings of subjects are important sources of data. Schools advertised in order to win students; therefore, presumably, advertisements reveal what subjects school leaders imagined students would find appealing.[38]

Virtually every educational institution, and virtually every educational theorist, listed English as the subject most important for all students to master. When Benjamin Rush outlined subjects appropriate for females, first on his list was a solid knowledge of English and the ability to read, speak, and write English correctly.[39] The study of English encompassed the acquisition of a range of skills: reading, writing, spelling, orthography, grammar, composition, speaking and oratory or rhetoric. Not every academy taught all of these branches, although nearly every academy taught reading, writing, and spelling to both sexes. Although these might seem like rudimentary subjects, they were not regarded as such at the time. Students in the all-male colleges also studied English grammar in the mid-eighteenth century. Princeton's curriculum, for instance, included spelling and grammar, and grammar was not relegated to the level of an entrance requirement until 1819. Similarly, the University of North Carolina maintained English as part of the formal curriculum until 1795, when it became an entrance requirement. Yale required Webster's English grammar text in the late eighteenth century, and Harvard did not reduce grammar to the status of an entrance requirement until 1866.[40]

Likewise, educators viewed orthography and handwriting as important for both sexes, a view that represented a change from earlier in the eighteenth century. Instructors considered spelling and orthography to be different skills, and schools did not necessarily teach them both. Students learned spelling by recitation; the teacher listened as students spelled words out loud. Orthography, however, was a written skill, and schools that emphasized reading over writing did not teach orthography. Colonial town schools, for instance, were more likely to teach girls to read than to write, so while some colonial girls learned spelling, fewer learned orthography.[41] Orthography was an advanced skill when compared to reading or spelling, and one that few females learned as a matter of course until the academy movement of the late eighteenth century. Handwriting also was a separate subject, and some "writing masters" advertised the various styles they were qualified to teach. Writing in a good clear hand was key to both good business and personal communication. Rush, for instance, in advocating this as an appropriate subject for females, discussed the importance of legibility and neatness in business. He opined that few things were "more rude or illiberal than to obtrude a letter upon a person of rank or business which cannot be easily read."[42]

In the 1780s, schools offered instruction in composition, or essay writing, less frequently than they taught reading, writing, and spelling, but this quickly changed. By the early nineteenth century, academies and seminaries more commonly offered studies in composition, or even required students to write essays. The initial lack of emphasis on composition for females is not surprising given that orthography, a foundation skill for composition, was just gaining in popularity for girls during the 1780s and 1790s. Composition apparently was not part of the original curriculum at Poor's Young Ladies' Academy of Philadelphia, but trustees added it in 1788.[43] Nor did Brown's Young Ladies' Academy of Philadelphia originally teach it, although one of the visitors, John Swanwick, appealed in 1787 for it to be taught.[44] Composition began to be taught in coeducational reading schools in 1789 in the Boston Public Schools, and appears in course listings at coeducational academies in the South in the 1790s.[45] One reason for the importance of writing was that letters were "lifelines" connecting women to the world and to distant family members.[46]

Educators thought speaking was an important skill for females as well as males. They did not expect women to make public orations, but they did want women to have proper diction, tone, and pronunciation. Not only would this help create, as Webster promoted, a distinctly American tongue, but it would also help distinguish people by class. Further, reading aloud to one's family in the evening was a task (or honor, depending on how one looked at it) that often fell to women. Being able to perform that task in a way that would please listeners was important. A trustee of a Philadelphia female academy encouraged students to learn to read aloud "with propriety and grace," and termed good reading a "charming accomplishment."[47]

Schools were less likely, on the other hand, to offer oratory to girls. For boys, rhetoric and speech making might be necessary for future roles as ministers, lawyers, or legislators. These positions were not open to girls, and so learning rhetoric in schools seemed less necessary. There were some exceptions to this curricular pattern. William Woodbridge apparently taught rhetoric at his female academy in New Haven in 1780.[48] Poor's Young Ladies' Academy of Philadelphia began offering a prize in rhetoric in 1790.[49] In 1792, the valedictorian of that academy justified girls' study of rhetoric, arguing that many subjects undertaken by boys had no specific future use either. For instance, argued Molly Wallace, most boys who learned Latin seldom used it. Most learning, she contended, was useful for the habits and discipline required, rather than for the content itself.[50] Franklin, too, advocated a study of rhetoric, not necessarily to learn how to make speeches, but to learn how to distinguish between emotion-stirring sophistry and well-reasoned argument.[51] To Franklin and many others, this seemed as important for women as for men, and rhetoric became common in female schools after 1800.[52]

These English studies were basic subjects, and, with some exceptions, by the early nineteenth century virtually every type of school taught them. After about 1840, when more students received rudimentary education in common schools, academies and seminaries listed spelling, reading, and writing only as preparatory studies, and taught grammar, composition, rhetoric, and belles-lettres as higher branches of English.[53]

After English, academies were most likely to advertise courses in arithmetic and geography. Virtually every academy—male, female, or coeducational—taught these subjects, although to what degree is difficult to determine. Arithmetic was important to women who would manage household accounts, and some academies specifically taught bookkeeping to girls. Benjamin Rush, for instance, believed that knowledge of figures and bookkeeping was "absolutely necessary" so that a woman could assist her husband or serve as executrix of his estate should she outlive him.[54] Benjamin Say, a trustee at Poor's Young Ladies' Academy of Philadelphia, urged students to apply themselves to arithmetic so that they could "buy or sell advantageously—cast up accounts, and in general [be able] to transact such business as may be found occasionally necessary."[55] Other academies, along with many district schools (public elementary or secondary schools), taught facets of arithmetic relevant to commercial transactions. Rowson's female academy in Massachusetts, around the turn of the nineteenth century, taught students how to compute interest, and also taught arithmetic through the mysterious "Rule of Three," which were rules for establishing relative values of various commodities.[56] Academies taught girls as well as boys, then, arithmetical skills useful in a market economy. At least some people saw this training as potentially vocational. The writer of a 1797 article on female education argued that there were a variety of trades and professions related to the practice of arithmetic and bookkeeping that should be opened to women.[57]

Virtually every school also taught geography. What was meant by geography, however, is difficult to discern. Kim Tolley concludes in her work on science education that geography entailed a wide array of subjects. In the late eighteenth century, many subjects were subsumed under the heading of geography, including history, physics, botany, geology, meteorology, medicine, and astronomy. Jedidiah Morse's widely used textbook *Elements of Geography*, for instance, included a history of science that introduced students to both ancient and modern figures in science, from Pythagoras to Boyle.[58] By the early nineteenth century, schools taught chemistry and physics as part of a course in natural philosophy; taught zoology, botany, and geology under the rubric of natural history; and offered separate courses in ancient, modern, and U.S. history. But in the late eighteenth century, all of these were variously taught under the label of geography.

Leaders of the new republic looked to education to unify and safeguard the nation, and geography textbook authors were willing to do their part to help. Texts extolled the beauties as well as the natural riches of the land itself, encouraging young people to feel pride in their country. Authors compared the people of other countries unfavorably to those of the United States. Morse, for instance, wrote that the Irish were a "blundering sort of people," the English were "proud and haughty," and Mexicans had "all the bad qualities of the Spaniards . . . without [their] courage, firmness and patience."[59] Textbooks also compared systems of government, with, of course, republics coming out far ahead of monarchies.[60] These lessons in national pride were intended for young people of both sexes. Morse dedicated his geography text, for instance, "to Young Masters and Misses Throughout the United States."[61]

Some academies offered modern and ancient languages. All-female schools frequently offered instruction in French. This was the case at Rowson's in Massachusetts, Brown's in Philadelphia, O'Connor's in South Carolina, and several female schools in North Carolina. The female Boarding School of Salem, North Carolina offered German in 1807. Some academies offered ancient languages, as well. Blair's private school for girls in the 1790s and Richmond Female Academy in the first decade of the nineteenth century taught girls Latin, Greek, and French.[62] The proprietor of the Cornelia Academy in Richmond, Virginia offered Latin and Greek to young women, believing that such study would offer "private enjoyment, social satisfaction, and permanent usefulness."[63] Coeducational schools in both the North and South offered Latin and Greek, such as Cummings' school in Duxbury, Massachusetts, where Sarah Ripley and other girls learned Latin at the turn of the nineteenth century.[64] A traveler in South Carolina noted that it was "fashionable for the young ladies [there] to be taught the dead languages [and] French and Italian."[65] Between 1810 and 1819, as many as 17 percent of female schools in North Carolina advertised that they taught Latin.[66] Academies far more often taught Latin and Greek to males than to females in this period, and some people vociferously objected to the idea of Latin and Greek for girls. "I know no way of rendering classical knowledge ridiculous," wrote the anonymous author of an article in *Lady's Magazine*, "as by cloathing [*sic*] it in petticoats."[67] But clearly some schools, as well as private tutors, did teach these languages to women, and the number of such schools increased during this period.

One reason for the increase may have been the changing meaning of classics. In the colonial era, Latin and Greek were the languages and the tools of ministers, doctors, and lawyers. People not entering those professions, which included all women and most men, did not need a classical education. But by the Revolutionary era, Latin represented far more than a dead language. It represented ancient Rome, a republic that the new American nation saw both as a model and as a warning. The story of the rise and fall of Rome had as its moral that republics are inherently fragile, and that their survival depends on the virtue of their citizens. Therefore, all citizens could benefit from learning about Rome. The early national period was, as historian Caroline Winterer phrases it, a "culture of classicism." Classical allusions abounded in newspaper and magazine articles and broadsheets, in public speeches and jokes, and in popular theater.[68] Understanding these references did not require that the average person be capable of reading classical texts in the original languages, of course. Even critics of male four-year colleges urged the adoption of reading these texts in English translation. But the pervasiveness of these allusions probably made it more generally acceptable for women to study Latin and Greek if they chose to. It was no longer a project tied solely to one's profession; rather, understanding classical culture was a project for all good republicans.

In addition to the academic curriculum of English, geography (broadly conceived), arithmetic and mathematics, and ancient and modern languages, academies customarily offered a number of nonacademic subjects, as well. One of these was needlework. This term encompassed several different sorts of work. Schools often specified whether they offered plain or fancy (or fine) needlework, or embroidery. Plain needlework was essential for nearly all females. As Mary Beth Norton notes, most women "devoted many hours each day to their needles," conducting the seemingly endless tasks of mending and altering clothes, and sewing shirts for their male relatives and caps, aprons, and dresses for themselves and their female relations.[69] Jeanne Boydston points out that even women in wealthy urban families did not escape the chores of needlework.[70]

Not only was needlework a practical skill that contributed to the domestic economy, it also had wage-earning implications, given that some women supported themselves or contributed to the family's income by sewing.[71] An article in the *New England Quarterly Magazine* urged young women to learn needlework, which the anonymous author termed an "accomplishment." But in this case, an "accomplishment" clearly was not meant to be merely for show, for the author went on to say that through needlework, women "may obtain an honest and honorable subsistence."[72] Others, too, advocated needlework as a way "the indigent may most speedily earn bread," or a middle-class "lady, in case of unexpected misfortunes, [may be guarded] from the horrors of dependence."[73]

Still, it certainly was not the case that academies universally offered instruction in needlework. Small schools with only a single male teacher did not offer needlework, including the well-known Northern schools for girls run by future Yale University president Timothy Dwight, or by William Woodbridge. Neither Brown's nor Poor's female academies in Philadelphia, both much larger than Dwight's and Woodbridge's, taught any form of needlework. Even when offered, some principals kept needlework in a clearly subordinate position. Sarah Pierce, for instance, did not permit students to enter her school solely to learn to embroider; they must also pursue academic subjects.[74]

Some academies in this period also offered music. Brown's Young Ladies' Academy of Philadelphia offered vocal music, no doubt influenced by the views of one of its trustees, the physician Benjamin Rush, who advocated singing and dancing for both boys and girls. Vocal music was accessible to all (unlike instrumental music, which required the expense of an instrument), and doctors credited it with several salutary effects. Rush believed that robust singing could defend the lungs from disease, and he asserted that singing civilized the mind and prepared it for the influence of religion and government.[75] Dancing, too, he viewed as promoting health, and not simply some frivolous pastime.

Educational theorists of the 1780s and 1790s did not endorse teaching instrumental music in schools. Rush discouraged instrumental music not only because of the expense, but also because playing well required too much time for practice, time that "could be better spent acquiring useful ideas," such as learning history or philosophy, or reading poetry or moral essays.[76] Maria Edgeworth argued that there was no real utility in learning to play an instrument, because after marriage few women had the leisure to devote to keeping up their musical skills.[77] In spite of the well-publicized views of Rush and Edgeworth, academies that taught music were more likely to offer instrumental than vocal music. However, only a minority of schools offered either before 1800. After 1800, a larger proportion of schools offered instrumental music, especially in the South.

Edgeworth recommended instruction in drawing rather than music. She suggested that fewer married women abandoned the practice of drawing as it "does not demand such an inordinate quantity of time to keep up the talent."[78] Drawing had a practical benefit related to needlework, as well. If a woman could draw her own patterns for her embroidery or sewing, she would not have to buy them or find someone to draw them for her. Even so, drawing was offered only in nine out of twenty-five schools in 1780–1810, and five of those nine were in schools that opened after 1800.

Other nonacademic subjects that were offered in the antebellum era, such as painting, lacework, and waxwork, were not offered in academies in the immediate post-Revolutionary decades. One might expect to see a move from less to more emphasis on academic subjects in curricula for women over time. Or one might expect to see an inverse relationship: that as schools added academic subjects or increased the rigor of study, they would drop nonacademic subjects from the curricula. In fact, institutions increased the numbers of both academic and nonacademic subjects over time. Schools added more courses in history, languages, and mathematics, but also added more listings in painting and music. The high rate of schools offering drawing held fairly constant. The exception was needlework, which was offered less often after 1830.[79]

The curricula at academies, then, exhibited little gender differentiation. Core subjects were the same for both sexes. Differences occurred most consistently in the area of vocational subjects: navigation and surveying for men, and needlework for women.

# Nobly Contending for the Prize: Pedagogical Practices

A wide range of writers echoed the belief that education consisted of far more than mere repetition of facts quickly learned and just as quickly forgotten. The ultimate goal of education, according to most of these authors, was to develop a person's reasoning powers, regardless of gender. Elizabeth Hamilton suggested that the function of education was to perfect a person's intellectual and moral powers. As noted in the quotation that opened this chapter, Hamilton saw no distinction between male and female needs for, or their abilities to develop, these powers.[80]

The standard mode of education in the colonial era, from primary schools through college, was that of rote memorization and student recitation. This remained the common practice in schools at all levels through much of the nineteenth century.[81] This pedagogic form had its opponents, however, many of whom spoke against its use for both males and females. In his 1798 plan for female education, John Hobson urged teachers to discern whether a student was a "wordy parrot or a reasoning machine." Science, he said, ought to be taught through Socratic dialogue. The point of education, Hobson insisted, was not merely the ability to recite a set of facts. Students must give proof of *real knowledge*.[82]

Writers and educators encouraged both male and female students to be active, not passive, learners. An article called "Hints on Reading" advised women to read systematically, analyze the books they read, examine an author's arguments, and "consider how far what he says concords with your own opinion and experience."[83] Similarly, Maria Edgeworth wanted women to be so confident of their own judgments that their "powers of reasoning [would be] unawed by authority."[84] Sarah Pierce, founder in 1791 of the successful school later known as the Litchfield Female Academy (attended by Catharine and Harriet Beecher), viewed genuine education as that which disciplined the mind. Pierce believed that mental discipline was equally necessary for girls and boys.[85] The active learning encouraged by these writers and educators—learning that would instill in young women a reliance on their own judgment—could not be a result of mere rote memorization. Instead, teachers asked students to explain the principles behind what they were learning, and not just repeat facts.

A second popular pedagogic tool of the late eighteenth century also encouraged aggressive learning in young women. This was the system of emulation—essentially, a form of competition, with prizes generally awarded in public. Teachers encouraged students to emulate the best student: to imitate with the intent to surpass. Emulation became controversial several decades later, as some people thought that it fostered the unchristian iniquities of envy and pride.[86] But in the late eighteenth and early nineteenth centuries, educators widely touted emulation as the best incentive to learning, for girls and boys alike.

Male visitors to and teachers at female academies praised the competitive spirit among the students. For instance, Reverend Sproat, a trustee of Poor's Young Ladies' Academy of Philadelphia commended students there who "have nobly disputed every inch of ground." Another trustee applauded their zeal, telling them, "You have nobly contended for the prize with very formidable and determined opponents, who disputed the ground with you, inch by inch, with praise-worthy perseverance and undaunted fortitude." In a graduation address, one student urged on younger students by saying, "With the spirit of enterprize [sic] and emulation, push forward your conquest."[87] John Swanwick, a trustee of Brown's Young Ladies' Academy of Philadelphia, referred to the "noble principle of emulation" and stated that this "sense of honor, this desire for fame, is a principle which can never be too much encouraged in every concern of life." Not only did he deem such a spirit of competition to be appropriate for women, but he also claimed that it was "implanted by Providence, no doubt, for very valuable purposes."[88]

Academy education in the 1780s and 1790s was very similar for men and women. Regardless of the sex of the students, most academies were day schools, not boarding schools. Educators applied to both boys and girls warnings about the potential of learning bad habits from other youth, as well as the perceived need for parental supervision of a student's nonclassroom hours. Academies offered almost exactly the same subjects to all students: the English branches, geography, and arithmetic, including that which was geared toward business.

Historians have pointed to two main subject areas as indicative of broad gender differences in schools: classics and ornamentals. But in neither case is the difference as significant as alleged. While schools did not routinely offer Latin and Greek to girls, neither did schools require these subjects of boys. Indeed, although many academies still taught the classics, educators disputed the importance of the "dead languages." In addition, girls studied the classical languages in coeducational academies as early as the 1780s and 1790s, and academies for girls frequently offered these courses in the early nineteenth century. Girls' schools' advertisements of "ornamental" subjects take on new meaning when seen in this broader context. "Ornamental" meant different things to different people, and sometimes meant anything beyond basic literacy. Also, some educators and parents promoted the ornamental studies of music, dance, and art for males as well as females.

The biggest difference in school subjects available to males and females related to possible future vocations. Many academies taught specific vocational subjects to boys, such as navigation or surveying. Similarly, many academies taught needlework to girls, either because they would need to know how to use a needle in their own homes, or because they potentially earn money with this skill. In addition, other subjects could lead to wage-earning positions. Solid English skills, including good handwriting, was useful in business, as was bookkeeping. Nonacademic skills, such as music and drawing, could lead to teaching jobs, as, of course, could the basic subjects of English, geography, and arithmetic. Decades before Catharine Beecher's famous appeal to make teaching an honorable and profitable profession for women, writers in the 1790s already were urging the same.[89] Some portion of the academy population, both male and female, attended school in order to have some vocational training. Some of that training was specific to one gender (navigation for boys and sewing for girls, for instance), while other training was not (English and arithmetic).

Academy instructors practiced the same pedagogical styles regardless of whether their students were male or female, and those styles were ones that promoted competition and the quest for "fame." Rather than attempting to socialize boys to be aggressive and girls to be demure, teachers apparently encouraged all students to succeed, and did so using martial language: students were to fight for academic prizes and to conquer the summit of knowledge. The ultimate purposes of education, that of teaching mental discipline and powers of discernment, were goals that educators had for girls as well as boys.

Opportunities for formal education increased dramatically between the mid-eighteenth century and the opening decades of the nineteenth century. While there still were some short-lived schools teaching one or two subjects, the "venture" school model faded out of existence and was replaced by permanent academies that offered a more full complement of courses under one roof. This expansion of educational opportunities for women continued in the antebellum era.

# Chapter 4

# Female Education and the Emergence of the "Middling Classes"

The popular novelist Maria Budden wrote in her 1826 *Thoughts on Domestic Education* that she was glad that fewer families insisted on teaching their daughters to play musical instruments. The "daughters of shopkeepers, and farmers, and poor gentry," Budden thought, needed more practical education than that.[1] The accuracy of Budden's comments was reflected in the growth in the number of educational institutions in the antebellum era, the curricula of these institutions, and the reality that the majority of these institutions drew people from the "middling classes."[2]

Historians have written extensively about the process of class formation in the 1820s and 1830s in the midst of an increasingly industrialized capitalist system.[3] The newly formed "middling class," which was never a clearly definable or monolithic group, included urban, nonmanual wage earners as well as farmers and "the moderate proprietor."[4] People in these middling ranks sought to distinguish themselves from those of both lower and higher socioeconomic status, projecting an image of themselves as the true progeny of the Revolution: hardworking, moral, Christian doers who were the backbone of the republic. Some of them found solace from tribulations and a sense of purpose in evangelical religion, and many, evangelical or not but fearing chaos in the midst of rapid changes in social and economic structures, joined temperance societies, moral reform organizations, and missionary groups. Seeking self-improvement, they formed mechanics institutes and Sunday School societies, and flocked to lectures at the lyceums.[5]

Education was central to the aspirations of the new "middling classes," and they contributed greatly to the growth of women's advanced education.

Evangelicals who emphasized the need to Christianize the nation, and especially the West, spurred the growth both of denominational colleges for men and seminaries for women. Advocates of the common school reform movement insisted that schools would preserve social order by inculcating the right values.[6] Their success, as measured by the increase in both the number of elementary schools and the number of students attending those schools, created a pressing need for teachers. Because of women's limited occupational options, and because of the evangelical spirit with which some women entered teaching, women were willing to work for little pay and to endure the uncertainties of irregular employment in the developing common school systems. Teaching young children was associated rhetorically with childrearing and therefore did not stretch conservative ideas about the roles of women too greatly, especially given that for many generations women had been teaching their own and other people's small children in their own homes in what were known as dame schools. As the common schools grew in number and importance, concerns about the quality of teaching led to a demand for adequate training of teachers, thus creating new opportunities for women in higher education. At the same time, the broad self-improvement movement of the antebellum era created a milieu in which both middle-class women and men actively sought education. Finally, faced with unstable economic conditions, women prepared to be financially responsible for themselves and possibly for their dependents. For many of them, preparation meant enrolling for a term or a year or two in a seminary, high school, or college. People pursuing each of these goals— Christianizing the nation, building a common school system, employing women as teachers, and women preparing for economic self-sufficiency— together contributed to the growth of women's advanced education.

In the 1820s and 1830s, a barrage of prescriptive literature rhetorically divided the world into female and male spheres. Denoted by some historians as a "cult of domesticity," the literature assigned women to the private world of the home, and associated the qualities of piety and morality with them.[7] Ideologies of separate spheres, however, did not mean that women were entirely constrained to their homes. In fact, middle-class women used these same ideologies as justifications for work outside their homes. If women were uniquely moral, as much of the prescriptive literature proclaimed, then they were specially qualified for benevolence work and moral reform.[8] Moreover, even those who prescribed limits on appropriate activities for women often regarded the realm of the intellect as suitable for both women and men. As a result, women's and men's advanced education was more similar than it was different. Excluded from most colleges, in institutions called seminaries, academies, and high schools, women received education very much like the education men received in the colleges, academies, and high schools they attended.

# A Happy and Holy Influence: Evangelicalism and Women's Duties

The great religious revivals that began around the turn of the nineteenth century and reached their height in the early 1830s were activist and millennialist. Many doctrinal differences separated evangelicals of various denominations, but what united them were their beliefs in a literal interpretation of the Bible, the need for spiritual rebirth, and acceptance of Jesus as one's personal Savior.[9] In the first few decades of the nineteenth century, some evangelical denominations, such as Methodists and Freewill Baptists, eschewed formally trained ministers in favor of those who simply felt "called" to minister. Believing that religious authority came from conversion, they disdained college-educated ministers and prided themselves on having ministers who were taught only by revelation. But by the 1830s, as these same denominations grew, leaders who had once prized an unlearned ministry began to call for an educated clergy.[10] The need for ministers, as well as the call for men to serve in foreign and domestic missionary movements, especially in the West, helped account for the rapid growth of denominational colleges, primarily for men, in the antebellum era.[11]

Rejecting a Calvinist belief that salvation was foreordained by God, evangelicals argued that not only could people work for their own salvation, they also could influence the salvation of others. Indeed, evangelicals believed that true Christians had an obligation to perform good works as one important manifestation of their salvation, and some of those good works must be attempts to convert nonbelievers so that they too could be saved.[12] The work that evangelicals envisioned was far greater than ministers and missionaries could handle, and was not to be engaged in just by professional ministers, but by all Christians. They did not leave this lay ministry work only in the hands of men. If they had, the work surely would have suffered, given that women outnumbered men both in church attendance and in conversion.[13] As the arbiters of sanctity, women had special responsibilities for improving the moral condition of society, thus ushering in God's kingdom.[14]

Evangelical Christians promoted advanced education for women based on their affirmation of a moral imperative to save an unenlightened world. Evangelicalism assumed women's moral superiority and their propensity for self-denying labor. Education, advocates argued, was a tool that would better prepare women for their work. In fact, evangelicals described learning itself as a religious duty for both women and men. "God requires all to take fast hold of instruction," Joseph Emerson told his Byfield Academy students in 1821, and Zilpah Grant and Mary Lyon repeated this command to their students at Ipswich a decade later.[15] Students in a Virginia seminary in

1825 learned that the "obligation to acquire true religion is compatible with the ardent pursuit of science and literature; acquisition of knowledge is expressly commanded by the Deity."[16] When understood in the light of a divine requirement, no subject was prohibited, because "[e]very science, every branch of knowledge, may promote the glory of the Divine Master."[17]

Not surprisingly, evangelical-supported schools featured moral education in their curricula. The purpose of education in the new schools that sprang up in the Northeast, the South, and the West was the "diffusion of literary and religious light through the world."[18] School founders and instructors viewed moral education as being critical for both men and women. In an 1835 address at the commencement of the exclusively male Hampden Sidney College in Virginia, James Garnett told students that true education "aims at perfecting our whole nature, intellectual, physical and moral," and objected to schools that instilled "the fear of man, the desire of applause and personal rivalry, in place of the fear and the love of God."[19] An advertisement for the male Amherst Academy promised that "[n]o pains shall be spared . . . to secure [students'] advancement in intellectual and moral culture."[20]

However, moral education took on heightened importance for women. "Moral excellence," said a writer in the *American Journal of Education*, "should be the great object of all human education; but this is peculiarly true in that of woman."[21] School circulars and catalogs frequently asserted the school's emphasis on moral education. A typical statement was present in the circular for the Adams Female Academy in New Hampshire, which stated the school's intent "to give in their highest possible degree, intellectual, physical, moral and religious instruction to Females."[22] After describing plans for intellectual and physical education, the Alabama Female Institute circular of 1835 stated "the conviction, that to the *moral* department of education, our most earnest and persevering efforts should be directed."[23]

Schools attended to moral education in various ways. At virtually all seminaries, academies, and colleges, trustees required presence at church services for male and female students.[24] In the curricula, courses in moral philosophy taught students to "reconcile . . . reason and natural law with . . . theology and Christian law." In most colleges for men, moral philosophy was a capstone course for seniors, usually taught by the college president.[25] Similarly, schools for young women invariably listed either moral philosophy or moral reasoning as courses for upper-level students.[26]

According to the rubric of evangelicalism, educated women had several arenas in which to perform their work. First, extending the Enlightenment and republican ideologies into the realm of Christianity, the literature asserted that women had a strong influence on their families—husbands, children, parents, siblings, and other members of their kinship networks. As one writer stated, "The moral power of woman extends not merely over children, but affects and directs the tastes, habits and pursuits of all her friends and companions. Her character is felt throughout the intricate machinery of society, and gives complexion to the age."[27] Those who believed that women held such sway over their families deemed education to be of paramount importance to ensure women used their power well.

Even more than in the early national period, rhetoric in the antebellum era emphasized women's roles as mothers.[28] Childrearing was religious work, according to many authors. Mothers were to "light the lamp of the soul, [and so] should know how to feed it with pure oil."[29] With such a weighty burden facing them, women needed all the intellectual and moral ballast they could get. The principal of a Kentucky female academy phrased the need for women's education in terms of spiritual warfare, saying, "Arm her with intelligence equal to the conflict; give her power, and knowledge is power."[30] Thus, belief in the importance of motherhood became a rationale for advanced education for women because "[w]hatever concerns the culture of the female mind, extends ultimately to the formation of all minds, at that early and susceptible period, when maternal influence is forming those impressions which eventually terminate in mental and moral habits."[31] Women's "arduous and responsible duties" for childrearing, according to one writer, were formidable, because "on the management of the child depends the character of the man, . . . by [woman] is formed the permanent, influential, moral, and mental character of mankind."[32] Writers frequently referred to mothers as the "first teachers" of the next generation.[33]

Many people articulated the belief that women also had an influence on society in general, outside their families. "What, then, are the stations, which females are called to fill?" asked Harvard-trained minister Joseph Emerson in 1822. "In these, there may, indeed, be some diversity," he answered. "All females have not the same office," and "there is not one station for all women," he said. While some women are called primarily to be daughters, sisters, friends, wives, and mothers, there are some who are also called to be teachers, "members of the church, of the civil community, and of the human family." A woman's influence "is not confined to her own dwelling," wrote Emerson, documenting women's work for "pious and benevolent societies, that are engaged for the improvement of the world."[34] Good education prepared women not only for their domestic lives, but also for lively engagement in the world. Emerson's insistence on this point is reflected in his choice to pursue a profession as proprietor of a school for young women; his most famous students, Zilpah Grant and Mary Lyon, went on to found successful schools of their own. The commonly held sentiment that women's influence was broad was captured in Abigail Mott's comments that the "safety and happiness of the nation depend on women," and in Samuel Burnside's assertion that the country needed "well educated women to uphold her institutions."[35] Elias Marks, principal of a Southern female school, wrote in 1828 that "if society should be happy and virtuous, women should be intelligent and virtuous."[36] In fact, wrote the female principal of a Northern school in 1833, "the moral atmosphere of the country depends on [woman]'s influence."[37] "The sphere of woman is, at the present day, far more extended than at any previous period," wrote Martha Hazeltine, a member of a female education society. A woman, she went on, "is no longer limited in her range to the sequestered walks of private life."[38]

Indeed, huge numbers of women contributed time, labor, and money to charitable works and to reform societies. Scores of women joined moral reform societies dedicated to eradicating sexual immorality. By 1839, the New York Female Moral Reform Society had 445 auxiliaries throughout New England, and it reorganized as a national society.[39] In Rochester, women founded the Female Missionary Society, the Female Charitable Society, the Female Association for the Relief of Orphan and Destitute Children, the Female Anti-Slavery Society, and a branch of the Female Moral Reform Society. They organized campaigns against alcohol, capital punishment, and tobacco, and spearheaded crusades for health reform and dress reform.[40] In the South, as well, women organized free schools for poor children, and ran societies that provided food, shelter, and alms for the poor. Southern women formed missionary societies and temperance organizations. Their work was serious and public, and was taken seriously by others, including state legislatures, to whom some female organizations appealed successfully for corporate charters. As corporate entities, women could, as did the Richmond, Virginia Female Humane Society, buy stock and use the dividends to fund their projects.[41] As middle- and upper-class women stepped out of their homes and engaged in benevolent works, the range of their influence increased to such an extent that one editor opined that "at no period of history, perhaps, have the sex exerted a holier or happier influence upon society."[42] Education could enhance this happy and holy influence; female seminaries both educated women for this work and encouraged them to undertake it.

A particularly notable example of how the seminary experience could encourage women to pursue benevolent work was that of Ipswich Female Seminary, where Emerson's two prize pupils, Zilpah Grant and Mary Lyon, taught in the early 1830s. A student at Ipswich wrote to her brother in 1831 about a series of lectures that "all seemed to aim at one point, to converge to one focus,—that of stimulating minds & hearts to action in promoting the happiness of mankind in general." The result, wrote Maria Cowles, was that "several are now willing to devote their time to teaching, who have hitherto been averse to it. Another is that several young ladies are willing to sacrifice home, friends, & New England privileges for the sake of doing good to minds in the Valley of the Mississippi." In addition, students organized a circulating periodical library, and established a system of dividing up articles among students who then would give summary reports. One rationale for this plan, according to Cowles, was that students and teachers alike regarded it as "important that young ladies should keep pace with the transactions of the day. They will feel more interest in promoting the grand object of doing good." Furthermore, those who participated would "learn the use of language, learn to arrange facts systematically & acquire confidence in speaking before such a company."[43]

Cowles' comments point to teaching as a third arena—in addition to having an influence on their own families and on society more generally—in which evangelicals believed educated women had the potential to Christianize the nation. Some reformers explicitly paralleled the spiritual work of ministers to that of teachers. Catharine Beecher is perhaps the best-known spokesperson for this view, which she first published in her 1835 *An Essay on the Education of Female Teachers*, but she was neither the first nor the only advocate of this viewpoint.[44] Ipswich Seminary had already been sending graduates out to teach in the West for years when Zilpah Grant wrote a public fundraising letter in 1836 asserting that few clergymen "are doing so much to promote the cause of education and religion" as these female teachers.[45] Evangelicals referred to women who went west to teach as "faithful laborers" who had "a strong missionary spirit" and acted from a "conviction of duty."[46] Men as well as women stressed the impact of women teachers. A circular of the newly founded Oberlin College stated in 1834 that its primary object was the "thorough qualification of Christian teachers, for pulpit and school."[47] Oberlin's statement is noteworthy for its implication of the equal importance of ministerial work and teaching. Evangelical Christians, both male and female, held female teachers in high esteem and believed them to be doing essential work in claiming the West for Protestant Christianity.

# "Then Send Women": The Feminization of Teaching

Evangelicals were not the only group encouraging women to enter the labor force as teachers. An increased demand for schooling, a low supply of men and a high supply of women willing to teach, and the belief that women were particularly suited for working with children all contributed to the shift from a teaching force that was mostly male in the colonial era to one that was predominantly female by the end of the nineteenth century.[48]

The movement toward increased schooling that began in the early national period became a veritable crusade in the antebellum era. "The spirit of inquiry, which has of late years extended to every thing connected with human improvement," declared an editor of the inaugural issue of the *American Journal of Education* in 1826, "has been directed with peculiar earnestness to the subject of education."[49] New York's governor Clinton stated that same year that the "first duty of government . . . is the encouragement of education."[50] Supporters of a system of uniform, public schools pushed a common school agenda while bemoaning the plethora of district schools and private academies.[51] Public or private, schools proliferated, and required a growing corps of teachers.

Men, however, were unwilling to fill the ranks of the teaching profession. In the growing economy, they had many more options for employment, and far more lucrative ones, that directed their attention away from teaching. As Catharine Beecher put it, "it is chimerical to expect, amidst the claims, and the honors and the profits of other professions, that a sufficient number of the male sex can be found, to devote themselves to self-denying, toilsome duties, for the scanty pittance allowed to our teachers."[52] An upstate New York minister and school founder wrote that "[s]o various and promising are the fields of usefulness, enterprise and ambition which lie before [men], few will engage in the work of instructing children and youth for a compensation such as the community are prepared to allow."[53] Teaching was a low-status occupation, and one that was also not as remunerative as other occupations open to men. The increase in schools and students, combined with men seeking work elsewhere, resulted in a demand for female teachers. A Western college president noted in 1833 that men used to teach but now were engaged in other ways. "Then send women," he concluded.[54]

A second circumstance motivated the new emphasis on women as teachers: schools routinely paid women far less for their work than they paid male teachers. In Massachusetts in the antebellum era, school boards paid women 60 percent less than they paid men.[55] By hiring women, financially strapped school districts could save money on their costliest budget item: salaries. Their desire—or need—to hire the least costly teachers further increased the demand for female teachers.

Lower wages or not, many women were quite willing to step in and fill the gap. For some, teaching resonated with their own missionary spirit. One applicant for admission to Oberlin College after another articulated her desire "to prepare myself for teaching in the destitute parts of the western valley," or to "qualify [my]self for teaching the poor blacks."[56] Far more men than women were migrating to the West, and New England in particular had a surplus of single young women in the early decades of the nineteenth century. Those women married later than the previous generation, or not at all, and needed ways to support themselves.[57] The factories of New England, along with other occupations such as milliner, seamstress, and domestic work, absorbed some of these women, but scores of others were available for teaching, and many more were able to do so than in earlier periods. Increased educational opportunities for females in the mid- and late eighteenth century meant that more females were qualified to teach. In Massachusetts, women comprised 56 percent of all teachers in 1834, which increased to 61 percent by 1840, and over 77 percent by 1860.[58]

The rhetorical and ideological associations of women with motherhood in the antebellum era also contributed to public acceptance of women as teachers. If, as Americans increasingly insisted, women were well suited for guarding the morality of their own children, it was a small step to further assert that women's natural abilities could be used to serve other people's children. "Nature has peculiarly formed and designed the softer sex for the noble and delightful, though arduous and trying, office of teaching," said one educator.[59] This logic could have backfired for proponents of higher learning for women. After all, if women were naturally suited for teaching, and destined to do so by Providence, why did they need special training? Seminaries, academies, high schools, and other institutions for women continued to increase. One explanation is that people understood that future teachers required higher learning and specialized training to perform their functions well. This explanation is supported by the importance of education in Enlightenment, republican, and evangelical thought, and the related trend toward credential requirements for teachers (which in turn made teaching even less attractive to men). Another explanation is that, while teacher preparation was one compelling argument for female education, it was not the only one. Indeed, women pursued education for reasons beyond evangelizing or preparing to be teachers.

## "Improve Every Moment": Women and Self-Improvement

"Improvement" was a catchword of the antebellum era. Everyone and everything was capable of improvement, and the fervor to improve was everywhere. "I delight in the spirit of improvement that characterizes the age," crowed an essayist in 1828. "I would not take away a single source of improvement. I would not close up an avenue to knowledge."[60] A profusion of new institutions sprang up to meet the needs of people from the middling classes and those in the working class who aspired to middle-class status. Mechanics' institutes provided libraries, reading rooms, and sometimes scientific equipment (usually cabinets of mineralogical specimens) to young working men who wanted to educate themselves. The lyceum movement drew people of all ages and various classes into lecture halls to hear discussions of current issues, scientific advances, or readings and analysis of literary works.[61]

Although the focus of historical writing on the self-improvement movement has been on men, women clearly were also a part of this movement.[62] Popular journals and literature, professional educators, and parents all exhorted women to improve themselves through education, and many women publicly and privately expressed their desire for self-improvement. "Every woman is invited to come to the fountain of instruction," wrote one male essayist. "The public feel it of infinite moment that each one should drink copiously. . . . Public light and feeling are demanding that women be extensively and thoroughly educated."[63] Another essayist referred to "the absolute obligation of every human being to improve every opportunity and use every means of mental improvement" in his argument for higher schooling for women.[64] The male principal of a female academy expressed his appreciation of his students' efforts, saying, "The exertions you are making for the improvement of your minds, the industry and zeal you have evinced in acquisition of knowledge . . . demand my gratitude."[65] Popular fiction reinforced this theme. A female character in an 1833 novel by Almira Phelps (sister of Emma Willard, the founder of Troy Female Seminary) tells her companions that she wants to learn because "it is right to improve ourselves in knowledge."[66]

Promoters of self-improvement spoke of it not only as being the right thing to do, but also as being a human right. In fact, one speaker contended that education was one of the "natural and inalienable rights of youth," comparable to a child's right to be fed. Furthermore, he argued, women had the same right to education as men, which was obvious now that the "cobweb notions about female capacity to acquire the higher studies have been swept away."[67] Another writer claimed that "mental improvement is the inalienable right of every being endowed with mind," and bemoaned the past in which "the mind of woman" was held back by "illiberal restrictions."[68] A male speaker celebrating the opening of two high schools for females in Worcester, Massachusetts said in no uncertain terms, "The human mind is formed to be cultivated. . . . God who gave us these faculties intended they should be cultivated, and women have the same right as men to enjoy the blessings of a cultivated mind."[69] The circular for Ballston Spa (New York) Female Seminary stated that women needed education "for [their] own rational amusement and usefulness," and to "contribute to [their] own gratification and improvement."[70]

Family members frequently entreated young people of both sexes to take advantage of their educational opportunities. "Youth is the season for improvement," counseled parent after parent.[71] "We wish you to always bear in mind that you go from home to *improve your mind*," wrote a Southern parent to a daughter who had gone north for schooling. The daughter responded in kind, saying, "I have made up my mind to study very hard. . . . I regret not improving my time better while at Hartford."[72] An older sister, already married and with children, wrote to a younger sister who was away at school that "you must endeavor to improve every moment of your time. Those are golden opportunities which if misspent will never return again."[73]

It was not just the young who desired to improve themselves, however. Adult women, including those already working as teachers, displayed a zeal for continued education. One such teacher instilled this value in her students, and continued to inspire them to pursue education long after they left her school. Martha Whiting wrote to a former student employed as a teacher, "Although you are now engaged in a school that does not require study to enable you to discharge your duties, yet be sure that you devote one hour or more every day to the pursuit of knowledge," apprising her that such study was "indispensable."[74] In 1829, Catharine Beecher attempted to recruit Mary Lyon to join her as a teacher at Beecher's Hartford, Connecticut school. In trying to lure Lyon away, Beecher listed the plentiful "opportunities to read and improve ourselves" available in Hartford.[75] Beecher assumed that opportunities for further learning would be a strong draw. Although Lyon did not take Beecher up on her offer, Lyon found multiple ways to continue her education, including taking a chemistry course at the Rensselaur Institute.[76]

Educators encouraged young women to see themselves as active agents in this quest for self-improvement. A lecturer at a Virginia school for girls urged students to take their schoolwork seriously, as it "is for your own benefit, your own reputation, and your own happiness."[77] The male head of a South Carolina female seminary cautioned his students that "[t]rue education is not the knowledge of things, but of self-cultivation of powers of understanding, correct reasoning, and correct acting."[78] A female applicant to Oberlin College bemoaned "the imperfect cultivation of [her] mental powers," and was encouraged when she heard of Oberlin, hoping that finally she might be able to enact her "cherished plan . . . to cultivate and bring into vigorous action all the powers of body and mind."[79] She believed that Oberlin was a place where teachers would help her achieve her plan of self-improvement.

## Preparing for the "Vicissitudes of Fortune"

Essayists spoke of improving oneself in moral tones, as a religious duty or a divine right. But they spoke of it in another way, one that had far more practical implications. Along with self-improvement, the middling classes of the early nineteenth century placed a premium on self-sufficiency, and did so for both women and men. Historians speak of the years following 1815 as the flowering of a "Market Revolution." The new scale of capitalist enterprises meant that some people became wealthy, while many more descended into poverty. In this environment of constant material instability and uncertainty, middling-class women needed to be able to provide for themselves.[80]

Writers frequently referred to the vagaries of the market, citing the "vicissitudes of fortune" that reduced "proud families to want." Speakers and essayists commonly used the phrase "reversals of fortune," and referred to the "uncertain events of this changing world."[81] Female factory workers in Lowell, Massachusetts wrote and published short stories explaining that country maidens and matrons came to work in the mills due to such reversals of fortune. "The Widow's Son," for example, tells the tale of a woman "reared in the midst of affluence" who, after her husband died, had to work as a seamstress to pay for the education of her son.[82] Similarly, in "Disasters Overcome," Sophia is the daughter of a wealthy farmer who suddenly loses everything. To help provide for her family, Sophia goes to work in the mills.[83] No one was immune to such potential life changes.

Educators and advocates of female education spoke to these fears, and offered solutions. A speaker at the opening of a high school for females in New York lamented that though sons were educated to be able "to buffet with the stream, and to put back or overcome the difficulties with which they may be surrounded," too little had been done for daughters. "How will they be enabled to struggle with those hardships, and meet those vicissitudes which they may experience in their progress through life?" he asked. It was "fallacious" to assume that women would always have someone to look after them. Far too many women have had to bear the burden of sustaining their entire families, and "it is the part of wisdom, to prepare [women for] those trials."[84]

Many people saw teaching as an occupation for which women should be trained. While evangelicals saw teaching as a way for women to save souls, those emphasizing female self-sufficiency spoke of teaching as a way for women to save themselves. Unemployed women were in grave danger, warned some writers, not merely of living in poverty, but of resorting to theft or even prostitution. Teaching was "one means of providing a respectable and useful occupation of females who . . . are dragging out a wretched existence, or [are] driven to the practice of crime by the want of adequate compensation, and often for want of employment."[85]

Although writers frequently repeated the theme that women were especially suited to teaching, essays also reflected a sense that women had few other options. Several writers noted "the want of professions adapted to the sex," especially since "manufactories which once furnished a large part of their domestic labors, leave many females, of all classes, without any useful occupation."[86] "The accustomed employments of women have been stolen away from them by the introduction of machinery," wrote a college president.[87] Joseph Emerson, a determined advocate of female education, raised the issue of teaching as a female profession at least as early as 1822, saying, "If the great business of enacting, explaining and executing the laws, and the still greater and more momentous business of preaching the everlasting gospel, are committed to men; if the lucrative and honorable professions of law and medicine are exercised by them exclusively, . . . does it not seem reasonable that the business of teaching should be in a greater measure consigned to the other sex?"[88] In 1826 Emerson stated his view that each woman should prepare herself to become a teacher, given that any and all women "may be necessitated in this way to earn their bread."[89] Catharine Beecher approved of what she saw as the throwing open of the teaching profession to women, offering, as it did, "influence, respectability and . . . the road to honourable independence [sic]."[90]

The possibility that there was a veritable army of unemployed women seeking remunerative positions was a potential threat to the social order. One writer noted in 1833 that "a spirit of impatience is manifesting itself among the female ranks" due to the lack of options for employment. "The offices of males are aspired after," he continued, warning that "it may be doubted if there be not danger of a civil war or worse." His solution was to make teaching a female profession, given that men no longer were willing to work for such low pay.[91]

Whether as a male, female, or mixed occupation, educational leaders wanted teaching to become a genuine profession.[92] For this to happen, the corps of teachers needed increased and more uniform training. In this way, the goals of the school reformers fit in well with the middle-class goal of self-improvement, and with women's desires both for economic self-sufficiency and for education itself. The general public had to be convinced, wrote one such leader in 1827, that "teachers must, like other classes of men [sic], be trained to their business." Teachers, he went on, have the power to "enlarge or narrow the public mind by the standard of their own education."[93] Parents and local school boards wanted teachers who were better prepared than the almost-cliché version of the old type of teacher: a man who was "drunken, foreign, and ignorant."[94] As historian Carl Kaestle points out, this image of schoolteachers may in fact have been created by school reformers pushing their agenda for professionalization.[95] Whatever the truth of the matter, by the early nineteenth century, parents and school officials began to raise their standards for teachers.

Before teacher preparation became more standardized and regulated with the establishment of state normal schools (a name taken from a French term for schools that offered systematized teacher training) in the late antebellum years, it was academies, seminaries, and high schools that provided teacher training.[96] Institutions of higher schooling designed special instruction for students who planned to become teachers as early as the 1820s. Adams Academy in New Hampshire was doing this by 1823.[97] When Boston's school committee established its High School for Girls in 1826, they adopted the monitorial system of instruction, which committee members thought of as "peculiarly appropriate" in the girls' high school "inasmuch as many, who will attend it, will not only acquire there an education for themselves, but will also learn the art of teaching others." The committee report went on to say that instructors in the city's primary schools were females, and that, "hereafter, many of them will be selected from those taught at the female High School."[98] The High School for Young Ladies at Greenfield, Massachusetts advertised the school's ability to instruct those who "designed to qualify themselves" to become teachers.[99] Men's schools also advertised special classes for those planning on teaching. Amherst Academy in Massachusetts did this as early as 1827, and by 1832, the academy comprised three departments—the classical, the Teachers, and the English.[100]

One aspect of teacher training was content-oriented. Simply put, a teacher needed to know algebra before she could teach it. In this sense, any school of higher learning prepared people to teach. A second aspect involved instruction not just in *what* to teach, but also in *how* to teach. Ipswich Female Academy advertised its intention to give instruction "on the manner of communicating knowledge to children and youth of different capacities . . . and on the manner of awakening attention, of exciting inquiry, of arousing the indolent."[101] In 1837, trustees changed the school's name to the Seminary for Female Teachers at Ipswich, indicating the extent of its popularity as a teacher-training institute.[102] By the mid- and late 1830s, the demand for teacher preparation programs grew, and school after school announced itself as being "designed for women to become teachers and educators of youth."[103]

Teachers themselves realized that they could do their work better if they entered the profession well prepared. Young people, and especially women, sought out schooling in order to become better teachers, even though there were as yet no degrees or certificates required. As cited earlier, for instance, letters of application for entry into Oberlin College were filled with the desire to be prepared to teach. In addition to women seeking preteaching training, those already teaching often came back to academies and seminaries for more instruction. Typical of teachers holding this attitude was Abiah Chapin, who wrote, "I have ever felt conscious of my own deficiency as a teacher," and applied for a scholarship to return to Ipswich Female Seminary for additional training.[104]

Several institutions and societies arranged for financial aid for students who intended to become teachers. Troy Female Seminary offered education "on credit" for such women, with the understanding that tuition would be paid later from their salaries.[105] Ipswich Female Academy put a similar system of loans in place informally in 1830, then formalized the system in 1835 with the formation of the Society for the Education of Females.[106] At its inception in 1839, Rutgers Female Institute offered free tuition and board to six young women each year who planned to enter the teaching profession and who had been recommended by the trustees of the local public school.[107] In 1838 Young Ladies' Association of the New-Hampton Female Seminary for the Promotion of Literature and Missions changed its name by adding "and Female Education" to its title. This name change reflected the society's new goal of providing scholarships for "indigent pious young ladies," so that such women's "ardent desire for usefulness [would not be] repressed by the hard hand of penury."[108] These financial aid programs opened up higher schooling to women who otherwise could not afford it. At the same time, such programs helped fill the teaching ranks with competent professionals, thereby promoting higher standards for teachers and potentially improving their status. In addition, financial aid programs provided opportunities for self-sufficiency for young women.

Advanced education prepared women for self-sufficiency in other occupations, too. In her 1839 book for teenage girls, *Means and Ends, or Self-Training*, renowned writer Catharine Sedgwick warned girls not to let themselves be rendered "helpless and dependent on men for support and protection," but instead be able "to secure an independent existence." She suggested several avenues of employment for young women to consider. In particular, Sedgwick recommended good handwriting as a means to earn a living, noting that writing was far better paid than sewing. In fact, she reported knowing a widow who supported herself and her two children by copying for lawyers. A second area of schooling she recommended pursuing with a mind toward self-sufficiency was arithmetic. "If women were well accomplished in arithmetic," she wrote, "many avenues to employments would be opened that are now shut against them," such as bookkeeping.[109]

Trustees of quite a few academies seem to have agreed with Sedgwick, as they included bookkeeping in their curricula for girls. Some catalogs specifically mentioned bookkeeping, such as that of Boston High School, which offered it to girls in 1826. Other schools listing this subject include St. Joseph's Academy for Young Ladies (Maryland) in 1832, Arcade Ladies' Institute (Rhode Island) in 1834, Newburgh Female Seminary (New York) in 1837, and Western Collegiate Institute for Young Ladies (Pennsylvania) in 1837.[110] Some schools might have included bookkeeping as part of their courses in arithmetic. Rochester Female Seminary (New York) in 1834 not only offered bookkeeping but also navigation, surveying, and mensuration, subjects more often associated with vocational training for boys.[111] Why Rochester Female Seminary did this is unclear. Was the seminary training women for these occupations, or were these subjects offered in the same spirit in which schools offered Latin—as teaching good mental discipline, regardless of any specific use to which such subjects might be put? In any event, it is clear that one object of higher schooling for females was preparation for economic independence.

The goal of economic independence for women dovetailed other aims of the newly forming middle class. The common school reform movement, supported by both evangelical and republican ideologies, sought to train young people to be productive workers and good citizens and to have the skills and fortitude to survive economic downturns. The school movement in turn both depended on women to staff the schools and to teach these lessons to children, and also created one important forum in which women themselves might attain economic self-sufficiency. Meanwhile, the cultural imperative for self-improvement reinforced the importance of education. In all these ways, education became a crucial means of both creating and sustaining the new middle class.

# Class Consciousness and Female Schooling

The expansion of higher schooling depended largely on increased numbers of those in the middling classes seeking education. These new ranks of students embodied a democratic spirit of claiming something rightfully their own that had once been denied them, and exhibited a degree of antagonism toward elites. Their goal in pursuing education was not to emulate the rich, but to create a new educated class true to republican ideals. In an 1826 speech delivered at a Phi Beta Kappa Society meeting in Boston, the U.S. Supreme Court justice (and future Harvard law professor) Joseph Story commented on the elitism of education in the past, and on the new, wider, dissemination of knowledge. Reading, he said, once was "the privilege of the few." It "required wealth to accumulate knowledge," and such learning "constituted the accomplishment of those in the higher orders of society, who had no relish for active employment." But now, in this "extraordinary age," knowledge "radiates in all directions; and exerts its central force more in the middle, than in any other class of society."[112] The implication was that while elites kept knowledge to themselves, and treated knowledge as a mere "accomplishment," the burgeoning middle classes seeking education would put it to better, and infinitely more practical and democratic, use. Story, whose father had been a member of the Boston "tea party," saw the democratization of education as central to the success of the republic.

Popular literature often portrayed wealthy people as not having enough sense to take advantage of education. In the 1833 novel *Caroline Westerley*, the middle-class, well-educated title character is knowledgeable, kind, humble, and, above all, sensible. This is in stark contrast to a wealthy young woman who complains about being forced by her parents to continue her education. She thought herself above the other girls in her school who, "after all, will have to get their living by their education."[113] In novels and stories like this, it is the young people of the middling classes who have a respect for education and who prepare for the "vicissitudes of fortune."[114]

Some determined democrats were suspicious of higher schooling, especially when a student had to board away from home. Advanced education, for both men and women, could be socially divisive. Sons and daughters might learn a foppish elitism that would cause them to disdain family members who had not gone on to school. One writer warned families of sending their sons to college, "vainly endeavouring to ingraft a classic taste or tone of honourable feeling" onto the family tree, only to meet with the sad result that the sons "acquire just enough of refinement in taste to despise the others."[115] An illustration of the ill effects of higher schooling appeared in the *Analectic Magazine* of 1820, which printed the engravings "Departure for a Boarding School," and "Return from Boarding School." The text accompanying these illustrations described a young woman who, before going away to school, wore "the unadorned and rustic simplicity" of a farmer's daughter, but who returned home "decked in the most fashionable attire" and expressing contempt for her family.[116] Advanced education might have opened up beyond the elite in society, but popular literature warned those in the middling classes not to take on airs along with their education.

Even as schools proliferated, they remained beyond the financial reach of many potential students. Mary Lyon, founder of Mount Holyoke Female Seminary, is well recognized for her desire to make education accessible to women of the middling classes.[117] In private letters she eloquently expressed her ache for affordable education. To her friend Hannah White she wrote, "My heart has so yearned over the adult female youth in the common walks of life, that it has sometimes seemed as if there were a fire, shut up in my bones."[118] To her mother she wrote, "My heart has longed to see many enjoying these privileges, who cannot for the want of means. I have longed to be permitted to [teach] where the expenses would be less than they are here [at Ipswich Female Seminary], so that more of our daughters could reap the fruits. Sometimes my heart has burned within me; and again I have bid it be quiet."[119] Lyon successfully made higher schooling accessible to a wider range of students through her "domestic system," in which students and teachers performed all of the cooking and cleaning, and even grew their own food.[120]

The same year that Mary Lyon poured out these private thoughts, an anonymous writer for *The American Quarterly Observer* made a public statement advocating higher schooling for middle-class girls. The writer spoke of the "large class of females in this country, belonging to families which are in moderate circumstances, who are debarred" from schools. A "farmer or mechanic, worth three or five thousand dollars," could not afford the $200 annual board and tuition demanded at most academies. As a result, "thousands, and tens of thousands" are excluded from getting "a superior education."[121]

One person who attempted to ameliorate this situation was the Reverend H. H. Kellogg, who in 1834 opened a Domestic Seminary for Young Ladies, which may have been one inspiration for Mary Lyon's domestic system at Mount Holyoke.[122] At Kellogg's school in Clinton, New York, students worked several hours each day in housekeeping, laundering and mending clothes, cooking, washing dishes, nursing, or sewing. Monitors kept records of how many hours each student spent working at these tasks, and the value of their time and work was deducted from their bills for tuition and board. Kellogg made a gendered argument for affordable education, insisting that women ought to be able to have "a thorough education, on terms as reasonable as are afforded to the other sex." He asked why "the widow's mite, the abundance of the rich, and the resources of the state, must all be put in requisition to educate young men? Are women so much more wealthy . . . that they can educate themselves?"[123]

Indeed, dozens of colleges in the 1830s tried various manual labor systems to defray costs for students who were not wealthy.[124] Male students at Oberlin produced food for the college community, while female students cooked, laundered, and sewed. The "grand object" of this system, proclaimed an 1834 circular for Oberlin, was to defray the costs of education.

The plan reduced expenses "to less than one half of what they are at most other Collegiate Institutions," thereby "extend[ing] the benefits of such education to both sexes and all classes of community."[125]

Other individuals, societies, and institutions made their own efforts to make schooling more affordable. As noted earlier, Troy and Ipswich Female Seminaries offered loans to students who promised to become teachers. Leaders of Troy congratulated themselves for this record of extending schooling to the needy. Almira Phelps declared that "it is the pride of this institution, that the daughter of the most humble mechanics and farmers, and of the wealthiest and most powerful of our citizens, here meet on terms of equality."[126] The backbone of Mary Lyon's successful fundraising was women of the middling classes. Rather than try to appeal to a few rich donors, Lyon literally went from door to door and town to town collecting dimes, dollars, and furnishings for her new school.[127] Some of her donors were self-sufficient women who gave what they could, including a 70-year-old woman who had "supported herself by her own exertions" and left $600 to Mount Holyoke at her death, and two "unmarried women who spin and weave for a living [who] gave $100 each."[128]

In this new world of higher schooling that encompassed students from a broader range of class backgrounds, some felt that education must be practical. Renowned educator Thomas Gallaudet advocated arithmetic, geometry, and algebra for females, but added that higher mathematics must be kept subordinate to arithmetical applications in a woman's daily life. Women should be able to keep accounts, for instance. He encouraged the study of composition, but railed against women who could write poetry but could not write a plain business letter.[129] Yet, as I show in chapter 5, "practicality" often was in the eye of the beholder, and not all middle-class students wanted purely practical courses.

It is in this context that some of the most disparaging comments were made about nonacademic subjects in female seminaries. For instance, Leonard Worcester, president of Newark Young Ladies' Institute, wanted to be clear that his school did not follow the "pernicious example" of schools for the wealthy, "where expenses are lavished on the ornamental branches, to the sad neglect" of more useful subjects. "A foolish emulation of the rich," he continued, "has often induced those in more dependent circumstances" to dismiss studies that might better prepare them for the future.[130] A young woman applying for admission to Oberlin in 1836 wrote that she had "felt unwilling to go to most of our Seminaries, where the great object is to make mere butterflies of women. I wish to go where not only the intellect, but the moral principles will be cultivated, disciplined, and trained for active service."[131]

# "A Trifling Away of Time" or
# "Elegant Accomplishments"

Given the emphasis on evangelicalism, self-improvement, and practical needs to prepare for financial independence, the ubiquity of courses in music and the fine arts in early schools seems surprising. Music and art comprised frequent offerings for females. Of 98 institutions for which such information is available, 66 offered music in the 1820s and 1830s, and 68 offered art. The proportion of academies and seminaries offering art held fairly constant, but the proportion offering music increased from 57 percent in the 1820s to 72 percent in the 1830s.[132]

Educators, students, and essayists offered a range of reasons to support the formal study of music and art. In some milieus, proficiency in one or more of the arts was a *sine qua non* for a woman entering fashionable society.[133] Some social circles highly valued adeptness in music, dance, or the arts for adding "charm and variety to social relations."[134] Another often-stated reason for pursuit of these subjects was one that held high currency in the early republic: women's need for safe and uplifting entertainment. Women would not have access to the same leisure activities and outside interests as men. Therefore, given that they must "look for their joys chiefly at home, they must have resources for domestic and lonely enjoyment."[135] A "cultivated taste," said the principal of one female academy, "is one of the most valuable resources under the trials and vexations of life."[136] Still others saw competence in various arts in a very practical light, as a means of earning a living.

On the other side of the ledger, some opposed these studies. Essayists and editorialists disdained the arts as remnants of an elitist society no longer appropriate in a republic, and criticized an overemphasis on these subjects in girls' schools at the expense of academic subjects. These critiques did not serve to eliminate training in these subjects, but they did force principals of some academies to defend having them in the curricula. Rhetoric regarding the proper role of "accomplishments" did not vary much by region. North and South, educators agreed that these subjects should be "kept subordinate, and sufficiently subordinate," to other studies.[137] The great defect in female education, said a speaker at the Female Academy in Albany (New York), was "that too much regard has been had to what is showy rather than what is solid."[138] An editorialist in 1835 railed against "that extreme in Female Education" that focused almost entirely on "the desire of display," and commented ruefully on "what a trifling away of time" such schools promoted.[139]

While seeking to avoid that "extreme," few female schools eschewed music and the fine arts altogether. An announcement for a female department of a public high school in Providence, Rhode Island said, "Far be it from us to undervalue ornamental education; we only wish it confined within appropriate bounds."[140] The Tuscaloosa Female Academy's 1832 catalog stated that the trustees "give no countenance to the acquisition of superficial accomplishments," but quickly reassured patrons that "it is not meant, that the ornamental branches of education shall be neglected; but that they shall not monopolize" the curriculum. The regular course included ancient and modern geography, history, geometry, algebra, moral and natural philosophy, political economy, botany, chemistry, astronomy, and Kames' Elements of Criticism. For extra fees, students also could learn fancy sewing, music, modern languages, drawing, and painting.[141]

Music and the fine arts almost always incurred significant extra fees beyond the regular tuition. For instance, at Day's Seminary in Boston in 1834, students paid $5.50 a term for instruction in the higher English branches, and $10 more per term for instruction on the piano forte. If students wanted to use the school's piano, they paid an additional $2 per term. The Alabama Female Institute charged $20 per five-month session for instruction in the advanced English branches, $25 for piano lessons, $5 for use of the piano, and $15 for lessons in painting and drawing.[142] This pricing system gave the implicit message that these accomplishments, not part of the regular course of studies, were not part of the definition of a well-educated woman. At the same time, the higher cost of these lessons also gave the message that certain skills were accessible only to the wealthy. From the point of view of the survival of any given academy, the extra charges were pivotal. Art and music lessons were income-generators for schools, with fees as much as four times the cost of regular tuition.[143] The costs to the schools were higher, too, of course, as they had to pay up-front costs for pianos and other instruments, along with art supplies. Still, it seems evident that schools made money on these lessons.

Some educators viewed the arts as pedagogically relevant. Joseph Emerson, for instance, referred to drawing as a "useful art, not an elegant accomplishment," and recommended having students draw pictures of famous figures as a way to interest students in history and biography. Likewise, he suggested that drawing maps would help students learn geography.[144] Emma Willard learned geometry by studying linear and perspective drawing, and thereafter taught her students in the same way.[145] This was often the case for male students, as well. For instance, Amenia Seminary in New York included "shades, shadows, and linear perspective" in its mathematics department.[146]

Arguments about pedagogical relevance did little to assuage those who believed that instruction in music and fine arts was inherently elitist and therefore inappropriate for students from the middling classes. Some writers

straightforwardly acknowledged the obvious fact that society was class strati-
fied, and that one's education should be in keeping with one's class. Maria
Budden, in the passage that opened this chapter, saluted those who "shall only
learn what is adapted to their stations" in life. In particular, she expressed her
pleasure that the once ubiquitous "attempt to acquire" proficiency in music
had been "laughed out of fashion." Skill in music required leisure time, she
argued, and those who, whether single or married, were likely to live on small
incomes would not have the time to practice. Teaching such girls music,
therefore, "is worse than useless, it is injurious, to waste time . . . [when] the
hours so devoted would be much more profitably spent in the attainment of
the useful branches of education—writing, arithmetic, and needlework."[147]
For Budden, then, music and drawing were accomplishments that only made
sense for a fashionable elite. Females of the middling classes needed to spend
their time learning more practical skills.

However, in a world in which rapid shifts in economic conditions
rendered class standing precarious, some educational reformers argued that
even the more fashionable accomplishments could be used in remunerative,
and therefore practical, ways. Almira Phelps made this point in her lectures to
young ladies, originally given in 1830 and subsequently printed in several editions. In these lectures, Phelps seemed to be responding to Budden's insistence
that girls should be educated according to their future stations. "Did we know
what would be the future situation of each one of you, we might proceed
somewhat differently in our efforts for your improvement," she wrote, "but
uncertain as are the events of this changing world," it would be impossible to
know what would be of the most use. Nor could anyone predict whether
those attainments would be "exercised only for the improvement and delight
of the social circle, or to be the means of gaining your own support and that
of others who may be dependent upon you."[148] Phelps here alluded to the
fact that proficiency in music, painting, or drawing might be a source of
family entertainment for some, but might be a source of employment for others. In light of the fees charged for instruction in these subjects, obtaining
sufficient skills to teach them to others could prove quite remunerative.

Of all the nonacademic subjects, needlework was the one that most easily
had practical applications, and yet a decreasing number of institutions offered
instruction in it in the antebellum era. Especially for women "in a middling
rank and with a moderate fortune," skill with a needle was imperative, as "a
considerable article in expense is saved by it."[149] However, even those women
from wealthier homes "who expect to entrust such affairs to the management
of others" needed to learn needlework in order to prepare for a possible
"reverse of circumstances."[150] Yet, while 52 percent of schools offered either
plain or fancy needlework in the 1820s, only 30 percent did in the 1830s. In
the 1830s, schools in the Northeast and West were far less likely to offer

needlework (16 and 27 percent, respectively) than schools in the mid-Atlantic or South (47 and 38 percent, respectively).[151] One reason for the decline over those decades may be related to the large increase in popular journals and magazines targeted to the middle-class housewife. Many of these magazines published needlework patterns and instructions for sewing projects that had not been as easily available before. With these illustrations and guides in print, women had less need for special tutoring from needlework teachers.[152]

The growing embrace of more equal educational opportunities for women in the antebellum years thus reflected many social currents in the new nation. Above all, the increase in educational institutions for women signaled the role that education would play in the construction and strengthening of the middle class. One segment of the middling classes— the segment that financed, built, and became students in hundreds of new seminaries in the antebellum era—was motivated by religious beliefs. Evangelicals used a gendered rationale to promote female education, describing women as more pure and moral than men. At the same time, though, the moral education they recommended was the same regardless of gender. Schools for both males and females emphasized moral education, and offered courses in intellectual philosophy, Scripture, and religion. Further, evangelicals saw both women and men as participating in the work of creating God's kingdom on earth.

The rise of the common school movement and the related increase in the numbers of female teachers also contributed to the growth in higher schooling for women in the 1820s and 1830s. The common school movement, too, was a reform that replicated the values and needs of the middle class, both because of what would be taught in the schools, and also because of the middle-class desires for self-improvement and economic self-sufficiency. The rhetoric used by the school reformers was highly gendered. Reformers asserted that just as women were naturally suited for the nurturing role of mother, so too were they naturally suited for the role of teacher. However, teaching continued to be a mixed-gender occupation. Women did not dominate the teaching force in any region until the 1840s, and in many regions they did not predominate for several more decades. A lower percentage of the teaching force was male in the 1830s than it had been, but teaching still was a common occupation for men, and both men and women sought training as common school teachers. So although much of the rationale for women to become teachers was different from that for men, the need for expanded education that would prepare them for their occupation was the same for both.

Some of the impulses toward expanded education for women clearly arose from less gender-driven concerns. The self-improvement movement of the middling classes in the antebellum era inspired both women and men to pursue education with assiduity. Volatile economic conditions made visible women's needs to prepare for economic self-sufficiency. Some of the occupations that women prepared for were also occupations for which men prepared: common school teaching; teaching painting, drawing, and music; bookkeeping; and scribing. Women and men also prepared for gender-segregated occupations, and this, plus music and the fine arts, were the arenas in which education was most marked by gender difference.

Finally, the class issues that affected the world of higher education affected it for both men and women. People made appeals for democratic as opposed to elite education on behalf of both women and men, and warned both about the dangers of "taking on airs" if their education were too fancy. Middle-class needs for financial aid applied to both women and men, as did opportunities to make school affordable, such as experiments in manual labor institutions.

Neither practical needs for self-sufficiency nor religious zeal to Christianize the nation fully explain why so many women avidly pursued advanced education. The motivations of women who sought higher learning were diverse and included educational desires that crossed class boundaries.

# Chapter 5

# "Perfecting Our Whole Nature"

## Intellectual and Physical Education for Women in the Antebellum Era

Women pursued advanced education for many practical reasons. As important as these practical motives were in the increase of female education, they do not fully explain why so many women flocked to high schools, academies, seminaries, and the few colleges that admitted them. The thousands of women who saw teaching as a way to earn a living, and the somewhat smaller number who prepared for other possible remunerative occupations as well, such as bookkeeping, scribing, or writing for publication, were not the only women seeking higher education, nor were future vocations their only goals. Beliefs in evangelical Christianity inspired women to become more educated so that they could be better moral influences on their families and the world at large, while concomitant ideals of self-improvement also motivated women to seek out formal and informal sources of education. Women pursued advanced education, not only for practical purposes, but also because they and their parents in a fundamental way valued learning for its own sake. The idea of self-improvement through intellectual growth was an assertion of these women's claim to their worth and independence in the Enlightenment, republican, and evangelical traditions.

In accounts of education in the antebellum era, historians have emphasized ideologies of intellectual difference between men and women. Those ideologies certainly existed. There were those who believed that men's and women's brains were not capable of the same feats, or that, even if they were, there still was no reason to educate men and women in similar ways.

Historians tell the stories of a handful of pioneers who challenged these beliefs, most notably Emma Willard, Catharine Beecher, and Mary Lyon. These women were indeed remarkable in many ways, and deserve recognition for a host of reasons, including the establishment of long-lasting institutions. A broader reading of primary documents suggests, however, that their beliefs about women's intellectual capabilities were not anomalous. These women were part of a large group of educators and advocates who believed in intellectual equality, that the pure pleasure of learning should be enjoyed by all, and that women deserved unbounded access to advanced education as a source of delight. This chapter examines these common views of women's intellectual capabilities, and their manifestations in the academic curricula available to women.

A second source of the conviction that education was intrinsically valuable to women was the common association of intellectual well-being with physical health. Educators and doctors filled journals with injunctions to exercise, and emphasized the importance of not building one's brain to the exclusion of building one's body. These directives applied to both males and females. Educators encouraged all young people to think of their bodies and minds as linked, and urged youth to develop both intellectually and physically. Like intellectual development, physical well-being served societal, religious, and individual ends. This chapter addresses the academic curricula as well as the availability of physical education in high schools, academies, and seminaries in the 1820s and 1830s.

# "Equals as Well as Friends": Male and Female Intellectual Capabilities

In a eulogy for Joseph Emerson, an early advocate of female education, an anonymous memorialist wrote that Emerson sought to "do away with the assumption that women were never designed to be literary or scientific." He "treated men and women essentially in the same manner, without any needless distinction," and he regarded women "as equals as well as friends."[1] Speaking in the same spirit as the advocates of companionate marriage in the previous decades, Emerson was not alone in his beliefs or in his work on behalf of female education.

Authors of numerous essays and addresses throughout the 1820s and 1830s asserted that women's intellectual capacities were similar to that of men's. For instance, the anonymous author of an 1826 article on women's education began by stating as a common fact that old ideas had passed away. "We happily do not live in an age," the author wrote, "when it is necessary to prove either the importance of education, or the propriety of extending it to females. The days are past, when a knowledge of tent-stitch, and the composition of a pudding or cordial was esteemed the chief glory of [women]." Instead, "[s]cience allures [them] to her temple," and they have become "the dignified and enlightened daughters of the greatest republic on earth."[2] For this author, the educational accomplishments of females reflected republican ideals. In his address at the Portsmouth Lyceum in 1827, Charles Burroughs claimed that women possessed "intellectual capacities and powers as great, and susceptible of as high culture, as those of men," and that "intellectual, physical, moral and religious education should be given in their highest possible perfection to woman."[3] An 1834 essayist made the same point, stating emphatically that "[t]here is no incapacity in the female mind for exertion in the highest departments of literature and science."[4]

So vociferously did so many proponents of female education assert women's high capacity for intellectual achievement that an occasional writer felt the need to play the devil's advocate. "It has been thought necessary to say so much about the intellectual equality of the sexes," wrote someone identified only by the initials E.N.Q., "that the fact is sometimes lost sight of, that there is a great difference between the minds of the sexes." This writer went on to argue that while all minds seek intellectual truth, males pursue truth through intellectual force and females through intellectual beauty.[5] What is notable in this example is not that someone argued that male and female minds had essential differences, but that this point of view was so rarely expressed. E.N.Q. certainly felt that most of the conversation about mental capacity was on the side of sameness, not difference.[6]

Indeed, writer after writer urged equal attention for male and female education. Believing, as so many did, in the similarity of men's and women's intellectual capacities, Charles Burroughs argued that "[t]here should be as much done for female children, as is done for our youth at the universities."[7] In 1831, leaders of Adams Female Academy in New Hampshire similarly asserted that "there should be as much done for the intellectual improvement of young ladies as is done for the youth at our colleges."[8] An 1831 essay echoed this theme, stating clearly that the "education of the sexes must be correspondent and coequal."[9] A few years later a writer for the *New York Messenger* took that city's public schools to task because they "exhibited a marked inequality between the sexes."[10] In 1838 educator and editor William A. Alcott cheered the "unequivocally successful" experiment of coeducation at Oberlin Collegiate Institute, where both women and men participated in "a course of liberal study." "The benefits which are likely to flow from it are immense," wrote Alcott. "Woman is to be free. The hour of her emancipation is at hand. Daughters of America, rejoice!"[11]

The reason behind the demand for equal education, according to many authors, was not that men and women would fill similar roles in life, but simply that both possessed intellects and both ought to have opportunities for exerting them. Often this position was stated in religious terms. "The Deity," one author wrote, "gave us not capacities for noble and lofty objects, to be denied their proper exercise."[12] Another expression of this theme came from Leonard Worcester, principal of a female institute, who wrote, "There is no good reason why [women] should not enjoy equal advantage for acquiring an extensive and thorough education. . . . Surely the God who gave us these faculties intended they should be cultivated, and females have the same right as the other sex to enjoy the blessings of a cultivated mind."[13] An anonymous essayist whose article was published in the *American Journal of Education* queried, "Does not the God of nature by endowing woman with the godlike faculty of reason, show us that it should be improved to the utmost?"[14]

One of the blessings of a cultivated mind, according to an 1835 speaker at the University of Georgia, was a heightened sense of self. "A consciousness of intellectual power," said Daniel Chandler, "would engender in every bosom a feeling of self-respect." Chandler saw this source of self-respect as being as pertinent for women as for men. "Give the female, the same advantages of instruction with the male; afford her the same opportunities for improvement," he urged, finally beseeching his listeners, "Shall not these advantages be afforded her?"[15] In a closely related vein, a speaker before a New England Young Ladies' Literary Society in 1836 explicitly included women in Enlightenment conceptions of humanity ideals by grouping male and female mental capabilities together and contrasting them with the mental abilities of animals. Rather than seeing a distinction between male and female minds, he saw them as similar, and saw "mind" as part of the definition of what it was to be human. "To say that mind is the standard of man," he said, "is to repeat a maxim equally applicable to woman: indeed, whatever there is of excellence or superiority in either, compared with the rest of animate creation, exists in that part of us termed intellectual; and in proportion as this is left uncultivated, do we approximate to the beasts that perish."[16] Therefore, cultivation of mental powers was essential for all humans, both women and men.

One logical extension of this belief was that the whole community benefited from advanced education for all. If education was what separated civilized humankind from wild beasts, then surely as many people as possible should be educated. The writer of an 1834 essay in the popular magazine *The American Quarterly Observer* advanced this point, saying, "Give to all men and to all women the best education possible. Cultivate their understanding in the highest degree. Our whole community ought to be raised upon a higher level."[17] Similarly, in Daniel Chandler's passionate argument for advanced education for women, he rhetorically established a chain that linked liberty to virtue, virtue to intelligence, and intelligence to education. He concluded that "our country itself will feel, through all of its diversified relations, the saving effects of the intellectual powers of the female mind."[18] The self-improvement movement that swept the nation in these decades clearly included women.

The rhetoric advocating increased educational opportunities for women stemmed from the reality that women did not have the same advantages of education as men. Some women expressed their frustration bitterly. While acknowledging that the situation had improved, Martha Hazeltine, Corresponding Secretary for Young Ladies' Association of the New-Hampton Female Seminary, wrote that it was difficult not to feel "deeply wounded." "We forbear to aggravate our resentment," she wrote, by enumerating the institutions, "munificently endowed, established for the education of men," or "place in mortifying contrast the utter destitution of such establishments for females." Hazeltine went on to argue that men and women have similar educational needs, and that if "the education of young men requires large and expensive establishments of halls, cabinets, libraries, apparatus, and a host of professors," clearly the education of young women required those same things. After all, women could not learn "by a sort of magic peculiar to ourselves." Men and women learned in identical ways, and so, she demanded, "give us the facilities for education enjoyed by the other sex."[19] Emma Willard, founder of the famous Troy Female Academy, also expressed her anger at the differences in opportunities for women and men. From 1814 to 1818 she taught in a small Vermont school for females that was right across the street from a college for males. "I bitterly felt the disparity between the facilities of the two sexes," she wrote, "and was led to a deep and constantly painful sense of the injustice which my sex had ever suffered in the inferiority of their advantages for mental culture." This experience became the impetus for her crusade to legislate state support for schools for women.[20]

Many advocates supported a movement toward increased financial support, whether that support came from state or local government, philanthropists, religious bodies, local donors, or other sources. To many, this was a self-evident need. One of the first and most famous attempts to secure public funding for schools for women was Emma Willard's address to the New York state legislature in 1819, which may have been partly responsible for female institutions receiving a share of the state's Regent-administered literary fund.[21] Catharine Beecher queried in 1827, "If the public sentiment has advanced so much on the subject of female culture, that a course of study very similar to that pursued by young men in our public institutions, is demanded for young ladies . . . should not the public afford facilities somewhat similar for accomplishing it?"[22] The editors of the *American Annals of Education and Instruction* asked rhetorically, "Can there be any good reason given why colleges or seminaries of any kind for educating young men should be endowed, which would not be equally strong in favor of endowing our female seminaries, and rendering them permanent?"[23] Although women did not receive equal funding for their advanced education, either from public or private sources, they did make strides forward during these decades. As noted, the New York state legislature included women's schools in its literary fund in 1819, and individual and denominational support for female seminaries helped found and sustain hundreds of institutions in the decades before the Civil War.

## "Elevating the System of Female Education"

The catalogs and reports of scores of female educational institutions make clear that many people believed that men's and women's educations should be similar. Historians generally have pointed to three seminaries as being in the forefront of this move toward similar curricula: Troy (New York) Female Seminary, founded by Emma Willard in 1821; Hartford (Connecticut) Female Seminary, founded by Catharine Beecher in 1832; and Mount Holyoke (Massachusetts) Female Seminary, founded by Mary Lyon in 1837.[24] While each of these schools did indeed offer academically rigorous educational programs, they were not atypical. Rather, they were prominent exemplars of a much larger move toward raising academic standards for females.

Numerous educators throughout the 1820s and 1830s called for reform of female education. They wanted girls to stay in school longer and to study more rigorously and in greater depth. As early as 1818, Joseph Emerson "set himself systematically to the great enterprise of reforming and elevating the system of female education."[25] The catch-phrases of the 1820s, used by Emerson and many others, were the need for a "systematic order of subjects" rather than the haphazard approach that previous schools had taken, and the need for "thorough" grounding at each stage in an orderly progression of study. Emerson referred to the "gradation of branches" necessary for a solid education.[26] In 1826, trustees of Massachusetts' Female Classical Seminary announced their intention to "remedy in some measure the evils which must necessarily attend that *superficial* course of instruction which has too long been pursued," and declared that "no pains shall be spared to make their whole course of instruction *systematic* and *thorough*."[27] Similarly, a Committee on a System of Education for the Albany Female Seminary recommended in 1828 a systematic and graduated course of study.[28] Instructors at New Haven Young Ladies' Institute promised a course of instruction that was both "thorough" and "systematic,"[29] while Boston's Young Ladies' High School promised "a thorough and extensive course of instruction," composed of "a regular succession of studies, systematically arranged."[30] Proprietors of Tuscaloosa Female Academy urged parents to permit their daughters to pursue the "full and systematic course," one that taught subjects "by a regular gradation."[31] When Bradford Female Academy was founded in 1828, the trustees had as one objective "to raise the standard of female education," and by the mid-1830s, the "thorough and systematic course of female education" taught at Bradford, as well at many other institutions, resembled courses available to males.[32] Leaders of Mount Vernon Female Seminary declared in 1836 that "schools of a decidedly high intellectual order should be sustained, in which young ladies may enjoy advantages similar to those afforded our young men in a course of liberal education."[33] The concern was not limited to individuals directly involved in female education. Stockholders of New York city's high school for boys, confident in the success of that school, "were anxious that a similar institution should be provided for Females," and resolved "unanimously" to buy land and build a three-story brick building.[34]

Heads of schools frequently asserted that, even though their institutions did not take on the name "college," they offered college-level curricula for women. For instance, the 1826 catalog for Newark Institute for Young Ladies announced that "a new era has now commenced: Under the present system, a regular collegiate course is pursued; Professors of Languages and of the Sciences of Natural and Experimental Philosophy, Chemistry, Botany, &c. are procured, and nothing [is] neglected."[35] Trustees of Brooklyn Collegiate Institute for Young Ladies designed it, according to its 1830 catalog, "to afford facilities to females in acquiring an education, corresponding with those enjoyed by the other sex in our colleges."[36] In 1830 the newly established New Haven Young Ladies' Institute intended "to afford to young ladies the means of acquiring a[n] education . . . not inferior in value, to what may be gained by the other sex in our High Schools and Colleges." The Institute offered to women the same program deemed "most efficacious in developing the mental powers of the other sex."[37] Trustees of a Massachusetts high school for girls promised that those students who stayed for the "full course" would make "the same acquisitions as are embraced in a Collegiate education."[38]

One of the most frequently given reasons for males and females studying the same or similar subjects was the need to develop mental discipline. "The cultivation of the *Reasoning Powers* [is] the primary object of all education," stated the author of one report.[39] "The true end of education," said a male speaker at a Southern female academy, "is to develope [sic], to unfold, to strengthen, to expand, to train and direct aright, all the intellectual faculties."[40] Men's colleges had been experimenting with the curriculum in the 1820s by allowing students to choose modern over classical languages, or to substitute science for the study of Greek. Predominating over these changes, however, was a belief in the need for mental discipline. No less an authority than Jeremiah Day, the president of Yale College, asserted that the primary purpose of higher education was to teach mental discipline through the "vigorous, and steady, and systematic effort" of in-depth study.[41]

Educators saw mental development and discipline as imperative for women as well as for men. Elias Marks told his female students in South Carolina that education consists "in the right ordering and training of the minds of youth, so as to impart a habit of correct reasoning, and a method of pursuing knowledge."[42] "Discipline of mind should unquestionably be the first object in all education," wrote trustees of a New Hampshire female academy in its 1831 catalog. "Females as well as males," they continued, "should be taught to think steadily and accurately and strongly."[43] Joseph Emerson clearly found mental discipline important, and apparently excelled in instilling it in his students. "He possessed uncommon skill in exciting those around him to *think*," wrote his brother, and added that this was "a habit which he deemed it especially important to cultivate in the education of females."[44] Students who learned good mental discipline would be able to apply the skill in any arena, bringing "its various powers to act on any incidental subject."[45]

Writers spoke of women applying mental discipline to their work in "forming and regulating the varied tempers, appetites, and passions of children and domestics, and for acquiring a decision and firmness, in the discharge of duties."[46] Members of Illinois Ladies' Association for the Education of Female Teachers concurred in the belief that mental discipline would help middle-class women in their daily lives. The Association's 1835 report stated, "The kind of education we would promote, prepares a person for severe application of mind, and for the correct and ready discharge of domestic duties. It enlarges, strengthens, and invigorates the mental powers. It teaches to reason, reflect and act."[47]

The belief that mental discipline was essential for women provided a rationale for the study of science, mathematics, and languages.[48] "Thus," said one advocate of female education in 1826, "those branches of science, which might at first view be pronounced useless to females, rise into importance from the habits of mental discipline which they establish."[49] When Providence High School opened a Female Department in 1828 that included courses in algebra, chemistry, and Latin, its trustees issued a statement explaining their rationale. "It may be asked, of what use will the study of Algebra, Geometry, &c. &c. be to young Ladies?" They answered their own question by saying that education "does not consist *merely* in storing the mind with knowledge. The mind must be *prepared* for the reception of this knowledge. . . . Hence the necessity of *mental discipline*, and such a course of study as will best tend to produce it. That Algebra and Geometry have this tendency, none will pretend to deny."[50] Similarly, the trustees of Boston's Young Ladies' High School wrote in 1831, "The question is some-times asked, of what *use* are Algebra, Geometry, Latin, &c. to females? Such studies are used rather as a means, than as the end, of education. They enable the pupil to learn other things with greater facility, from the salutary influence they exert upon the several faculties of the mind."[51] The president of Young Ladies' Association of the New-Hampton Female Seminary encouraged the study of classics "to exercise the memory, and increase the store of useful knowledge," and encouraged work in mathematics "to strengthen the reasoning powers, and prepare the mind for close thought, and rigid application."[52] Geometry, according to instructors at Adams Female Academy, "greatly invigorates and sharpens the understanding, by establishing habits of patient investigation, and of exact method . . . [It is] a kind of mental exercise and discipline, for which nothing is an adequate substitute," producing "a familiarity with abstract thought, a clearness of conception and precision of language."[53] The writer of an 1834 essay that appeared in the popular magazine *The American Quarterly Observer* stated that there was no reason that "a thorough *theoretical* education is any disad-vantage to females," and that in fact, "a thorough training of the mind in philosophy, mathematics, the languages, we mean the *dead* languages, will do more to qualify the female to discharge her duties" than any other sort of education would.[54]

In her 1826 treatise on domestic education, Maria Budden advocated mathematics in order to "obtain a better power of reasoning." Turning a stereotypical critique of women to her advantage, Budden suggested that women more than men would benefit from geometry. Women, she wrote, are "generally reprobated for the irrational facility" of gossiping and repeat-ing unsubstantiated rumors. Schooling in Euclid would solve this problem. As women developed "a predilection for proofs and demonstrations," they would be less likely to disseminate unfounded reports.[55]

Many educators believed that the study of modern and ancient lan-guages were especially fruitful ways to learn mental discipline. The head of an unnamed female school wrote that translating from Latin and French "has been the chief means" used in the school to foster the "power of form-ing clear, distinct, and accurate *Conceptions*—one of the most valuable attainments of which the mind is capable."[56] The writer of an 1827 treatise on female education believed that the study of ancient languages was "advantageous to every mind; it refines and elevates the tenor of thought and expression," it "imparts accuracy and gracefulness of style," it facilitates a better understanding of English, and therefore "is as valuable to the female sex as to the male."[57] Instructors at New Haven Young Ladies' Institute attached "great importance" to the study of ancient languages because mastery of those subjects "have, with scarcely an exception, succeeded best in every department of education," owing no doubt to the "vigor of intellect and power of application" acquired by those means.[58]

Even those few proprietors of schools that stressed the domestic nature of women's education still advocated education in science and languages, while virtually no schools taught courses related to housewifery or childrearing. For instance, an 1837 catalog for St. Mary's Hall, a female school in New Jersey, insisted that "*the education of females should be*, as nearly as possible, *domestic*," and emphasized that this was, in fact, "A SCHOOL FOR MOTHERS." St. Mary's Hall claimed to specially prepare young women for that exacting role, and the catalog listed a course called "domestic economy." But the bulk of the subjects listed in the three-year course for these future mothers, and which the trustees labeled "useful," comprised "every department of science, literature, and the fine arts," including natural philosophy, chemistry, botany, ancient languages, French, German, Italian, and Spanish.[59] Another school whose principal was clear in his belief regarding the domestic nature of women's sphere was the Ballston Spa (New York) Female Seminary. Students there did not learn geometry, algebra, or higher mathematics, as the principal believed that those subjects were only relevant for men. However, students took courses in natural and moral philosophy, history, arithmetic, bookkeeping, French, and Latin.[60] Similarly, the president of the society that established New York's high school for girls eschewed "abstruse branches of science," saying that such knowledge was neither necessary nor useful for females. Subjects he did find useful for future wives and mothers—subjects that would "prepare them for presiding with skill and prudence in those domestic stations, for which providence has designed them"—included astronomy, natural philosophy, history, bookkeeping, geometry, "higher arithmetic," geography, and modern languages.[61] Newburgh Female Seminary in New York listed domestic economy in its curriculum, along with other practical subjects such as bookkeeping and plain sewing. But these subjects certainly did not predominate, as the curriculum also included arithmetic, history, natural philosophy, geology, botany, zoology, algebra, and Latin and Greek.[62]

Those educators who had qualms about spending too much time studying the dead languages still advocated some degree of familiarity with them. "I do not hold the study of these languages in the highest estimation," wrote

Joseph Emerson, but "some acquaintance with them is really desirable, and should be possessed by every Young Lady."[63] Consistently, Emerson wrote to the president of the all-male Amherst College promulgating his view that a "limited knowledge" of Latin and Greek should suffice for students there, as well.[64]

Judged on the basis of course offerings, academic standards rose consistently in female seminaries and academies. Out of 91 schools for which such information was collected, 19 percent offered algebra in the 1820s, while 67 percent did so in the 1830s. Similarly, 34 percent offered geometry in the 1820s, compared to 74 percent in the 1830s. The percentage of schools offering botany rose from 38 in the 1820s to 72 in the 1830s. Chemistry, astronomy, and natural philosophy were popular courses in the 1820s and remained so in the 1830s. In the 1820s, 63 percent of schools offered chemistry, and the same percentage offered astronomy. In the 1830s, the percentages were 77 and 68, respectively. The percentage of schools offering natural philosophy stayed constant at 84 throughout both decades. Very few subjects reveal regional differences. The few differences that are apparent were in math more than science. In the 1830s, half of Southern and Mid-Atlantic schools offered algebra, while 80 percent or more of New England and Western schools did so. At the same time, over 90 percent of New England schools offered geometry, while only two-thirds of schools in the other regions did so. New England schools were slightly less likely to offer botany, while Western and Mid-Atlantic schools were more likely to offer astronomy than were schools in either the South or in New England.[65]

The number and percentage of female schools offering ancient and modern languages also rose. Latin, Greek, and French all were offered more frequently in the 1830s than in the 1820s, while Italian and Spanish either held steady or dropped slightly. In the 1830s a few schools also offered German or Hebrew. The percentage of schools offering Latin rose from 59 in the 1820s to 80 in the 1830s; those offering French rose from 66 percent in the 1820s to almost 85 percent in the 1830s, and Greek rose from 17 percent to 40 percent. French and Latin were the most popularly offered languages. In the Mid-Atlantic and West, they were offered with equal frequency. In the South, students had access to studies in French in 93 percent of the schools, and to Latin in 73 percent of the schools. In New England, the relative popularity of these languages was reversed. Students could study French in 84 percent of the schools, and Latin in 92 percent of the schools.[66]

Theories of pedagogical methods also attest to similarities between female and male education. Educators in the antebellum era continued the movement begun in the early national period that shied away from passive learning and rote memorization. Derision for the practice of students' mere repetition of textbooks instead of performing exercises that fostered genuine understanding filled educational literature. As one writer put it, "It was formerly too much the custom to strengthen memory at the expense of understanding, by requiring long lessons verbatim, or more properly, parrot recitations."[67] The ubiquitous Catharine Beecher weighed in on this issue, prodding instructors to provide females with "thorough and substantial knowledge instead of that parrot learning which will soon pass away."[68] One instructor angrily denounced those who, in 1837, still relied on the old methods. "The teacher whose business consists in merely hearing recitations; in listening to the mechanical repetition of what is contained in books and as mechanically asking the questions he may find there . . . deserves not the name of teacher," John Howard stated emphatically. True teaching, he went on, resulted in the student knowing "the *why* and the *wherefore* of all he studies."[69]

Schools advertised their disavowal of the pedagogical method of rote learning and repetition. The South Carolina Female Institute's catalog of 1828 asserted the school's goal of "awakening the curiosity, improving the memory, and exercising the . . . reasoning powers" of students. Students were not just to memorize rules of grammar or geometry, but to understand "the reason of the rule." "The mind must not be suffered passively to grope its way from one lesson to another," reciting lessons in the "dull, monotonous chime of a cloisteral recluse," continued the school's "Plan of Instruction." Far from it! Contrasting the "dull–heavy–plodding" nature of rote learning with the "elasticity and spring of thought" of genuine engagement with learning, the "Plan" goes on to declare that "an Academy ought to be a literary gymnasium, in which the competitor, even if she fail in obtaining victory, is intellectually strengthened by the nature of the contest, in which she is engaged."[70]

Educators applied this rationale to women as well as to men. In an 1832 address on female education, Leonard Worcester, president of Newark Young Ladies' Institute, contended that, while it is easy for a student to commit a lesson to memory, "nothing is learned until it is understood," and such understanding required not just memorization, but "the exercise of thought, and reflection, and judgment."[71] Toward that end, faculty at Boston's female high school encouraged to "use their own words [in recita-tions] in preference to those of the text books."[72] In a similar vein, trustees of Adams Female Academy stated that their instructors were not satisfied with students' "mere repetition of the words of the author," and instead would hold "conversation and discussions" in order to "bring the mind of the scholar to bear directly on the subject." Further, they warned, a student would be expelled unless she "evince[d] a decided inclination vigorously to apply herself to study."[73] Likewise, students at the New Haven Young Ladies' Institute were informed that "the hill of science can never be ascended without persevering, personal effort," and were warned that "the Principal wishes it to be distinctly understood, that any pupil, who cannot be persuaded to act upon this theory, *cannot retain her connection with the seminary*."[74] Instructors at Roxbury Female School planned to teach students "*how to study* and *how to think*, and *how to exhibit the results with clearness, conciseness and force*." Scholars would develop "habits of profound research, intense meditation and accurate comparison," leaving nothing to be "received on the authority of assertion."[75] Similarly, teachers at Gothic Seminary would not regard a student's education as complete until she "possesses the power of abstraction, accuracy in discrimination, and a capacity for thorough, and successful investigation." Students would know how and where to educate themselves upon leaving the seminary, as they would possess "the ability to acquire and impart knowledge on any subject."[76]

Journal articles celebrated Pestalozzian methods of learning by observation and experiment instead of reliance on books, methods that were adopted by numerous schools for both males and females.[77] Elizabeth Female Academy in Mississippi advertised its practice of principles derived from Pestalozzi in its annual report to the trustees in 1826, a report that detailed the school's use of "the *inductive system*" of learning.[78] Troy Female Seminary, under the leadership of Emma Willard, also adopted Pestalozzian techniques.[79]

Educators encouraged young women to be active learners, and disparaged passivity in students. True education "demands something more, than sitting still and passively receiving . . . more, than idly and stupidly receiving that information which you can scarcely avoid," Joseph Emerson lectured his students in 1821. "You must be active," he commanded. "You must be up and doing." Using the metaphor of knowledge as a precious gem, Emerson instructed students that they "must *dig* to find it. You must seek it as silver; you must search for it as for hid treasure; seek it, as the hardy and industrious miner seeks the precious ore. . . . Your finest ardors must be kindled, your noblest faculties must be exerted to the utmost, to gain all the information in your power."[80] His protégées, Mary Lyon and Zilpah Grant, told their students at Ipswich in 1834 that they "must do more than to sit still and merely receive. . . . You must seek for knowledge . . . your mind cannot be improved without your activity."[81]

Emerson, Lyon, and Grant may have been especially impassioned, but their central theme was a common one. Again and again, educators urged female students to apply themselves assiduously. James Garnett counseled students in a Virginia female academy to take as their motto "*Juvat transcendere montes*—it delights me to surmount difficulties." If students wanted to "be either wise or learned," Garnett enjoined them, they "*must* be content to encounter some toil for such an inestimable blessing," and not be misled into believing that all they have to do is "remain entirely passive and merely to listen to the words addressed to them, without any effort." Do not, he importuned, "perform a part very little better, than so many empty casks, which receive all things that are poured into them." Garnett urged not only active learning, but also independence of thought and action. Calling it a "fatal mistake" to rely entirely upon one's teachers, he told students that "the greater part of the business of Education depends upon ourselves. . . . Learn to rely on your own powers."[82]

Teachers at Ipswich Female Academy encouraged their students who were being trained to become teachers themselves to learn methods that would inspire thinking on the parts of their students. For instance, they were told that the "class should be incited to notice the operations of their own minds, and observe the practical effect of the principles of mental Philosophy." Students also were required to give "original examples of the operations of the mental faculties, different kinds of association, reasoning &c."[83]

Many people hailed independence of thought as a hallmark of true education. The educational reformer Thomas Gallaudet defined the "great object" of education as "qualifying the pupil to think for herself; to be able to pursue her studies alone; to discover truths, and form conclusions, and establish opinions besides those with which her memory has been furnished, either by her books or her teachers."[84] The senior class at Ipswich Female Academy wrote a statement of characteristics of that academy. In that document in 1829, they described how students "are induced to examine subjects for themselves, nor do they mean to admit the assertions of any author, unless convinced by fair argument, or by their own reflection."[85] Instructors at Mount Vernon Female Seminary declined to use a single textbook for any given subject. Instead, "the pupils are led to bring to the subject which they are examining, all the light which they can obtain from various authors, that they may learn to compare testimony, balance opposing opinions, and look for truth with careful discrimination, instead of receiving implicitly the views" of a particular author.[86] Instructors at Rhode Island's Arcade Female Institute believed that an "essential constituent" of female education was for women to gain "the ability to investigate, and to confirm by the respective methods of proof."[87] Proprietors of Watson's private school wrote in their 1834 advertisement that the school's design was "to teach the pupils in such a manner as to call all their powers of mind into exercise, [and] to promote habits of active and independent thought."[88] At Alabama Female Institute, students were taught "not merely to communicate words and facts, but to cherish the spirit of enquiry, and form the habit of thought." Instructors there aimed "to induce every young Lady, so to read, and observe, and reason, as to ascertain facts, and frame conclusions and opinions for herself,—*to become, in truth, her own Teacher.*[89] Female students at Oberlin noted that they were "taught not only to fully appreciate the worth of an author, but to *think* for ourselves upon the various subjects brought before us." The students followed up that description with the statement, "and we do feel that this knowledge after which we are searching is of more value than the diamond which sparkles in the sands of India and the pearl in its ocean bed."[90]

By the antebellum years, educators held mixed views of the pedagogic tool of emulation, which had been popular in the early republic. The conflicts were evident in an 1828 article on female education that appeared in the *American Journal of Education*. When the author of the article proposed awarding honors to students for high achievements in various subjects, such as Latin or French, the editor of the journal added a footnote that asked, "May not undue ambition be excited in this way . . .?" Such a system would be "apt to give rise to *envy*, and . . . may produce *pride*."[91] Although this particular article was specifically about female education, the same debate occurred regarding male education. On one side, some expressed the view that "talent laboriously cultivated . . . [should] have its distinction," and that competition spurred people on to greater achievements.[92] Those who had cynical attitudes toward students were especially likely to favor emulation. "Man is naturally indolent," said one such educator, and the "first object" for an instructor "is to overcome this natural propensity for sloth."[93] On the other side, critics argued that emulation "deteriorates the character" and "produces an avaricious desire for marks of personal distinction."[94] Not surprisingly, many schools announced where they stood on the issue. Boston's high school for girls, for instance, stated clearly that "[n]o medals or pecuniary rewards are given," as the trustees had no desire to "appeal to mercenary motives."[95] An observer of Ipswich Female Seminary noticed that teachers there discouraged "all display of attainments, all trivial distinctions, all direct comparisons of one with another," and that therefore there was no rivalry.[96]

## Women's Own Desires: Love of Learning

If women's education was similar to men's, one explanation, too often overlooked in historiographical accounts, is that women wanted it that way. Women valued education for its own sake, for the delight of partaking in the learning process, for the heightened sense of self that came from accomplishing academic feats, and as a source of unending comfort and pleasure. Schools increasingly raised their academic standards for women because so many of their students wanted advanced education on a par with men's.

One woman who epitomized the desire for education was Lucy Stone, who later in life would become a vocal abolitionist and advocate for women's rights. She was one of nine children in a farming family in western Massachusetts. Stone vowed as a girl that she would get an education. She sat in on the lessons her father permitted for his sons to prepare them for college entrance, but he refused to spend the money on a book for her. Barefooted, she collected chestnuts and berries in the woods and sold them—not to buy herself a pair of shoes, but to buy herself a textbook. When she was 16, her father insisted that she take a job as a schoolteacher in order to help pay the debt on the family farm and to help pay her brothers' college tuition. As she taught, she continued her college preparatory studies on her own, giving most of her salary to her father but saving small amounts for herself as she could. Nine years later, at the age of 25, she finally was able to enroll as a student at Oberlin College.[97]

Women's prizing of education did not occur outside a cultural context. Scores of essayists and educators promulgated a view that held education as one of life's great treasures. "Knowledge is a precious gem," Joseph Emerson told his female students in 1821.[98] James Garnett's set of lectures, printed in several editions in the 1820s, began with the statement that happiness "is the universal aim of mankind," and is to be attained "from the pleasures of sense and intellect." Garnett invited students to "open for yourselves all those exhaustless treasures of knowledge that furnish the proper subjects upon which to exercise literary taste and scientifick [sic] talent." The "great leading principle of our lives," he continued, "should be the constant love and pursuit of knowledge and virtue."[99] Maria Budden included in her *Thoughts on Domestic Education* the precept that "knowledge is valuable enough to be desired for its own worth," regardless of either practical application or public applause.[100] Another proponent of advanced education for women compared the pleasures of learning to the "noisy mirth" and "vulgar hilarity" that sometimes get mistaken for true happiness. "The pleasures of a cultivated mind—in the exercise of its faculties, and the pursuit of knowledge—infinitely transcend in kind, variety, durability, and value, any thing that by the unthinking of the world, is termed 'pleasure.' The greatest possible happiness," she instructed, "is found, where the mind is continually exercised in the acquisition of knowledge."[101] Trustees of Mount Vernon Female School noted in their catalog in 1831 that discipline was maintained, not by reliance on fear of punishment, but by "the interest which the young mind may be led to take in the acquisition of knowledge," knowing, as they did, that "mature minds derive so much pleasure from the pursuit of study." In fact, the school had one of its central tenets the "pleasure of faithful and successful intellectual effort."[102] A commencement speaker at the University of Georgia urged thorough education for women, noting that "[r]eading would be an employment, study an amusement, contemplation the source of happiness, and reflection the rich treasury of many a blessing." Not only would education be a source of pleasure, according to this speaker, it also would be a source of self-esteem. "A consciousness of intellectual power," he said, "would engender in every bosom a feeling of self-respect."[103]

Female students shared this view of education. "God has not bestowed a greater privilege," wrote a student in 1824; "it is the comfort of life."[104] Edification occurred not only in classrooms, but at mealtimes, as well. "Our meals are so deliciously seasoned with profitable conversation that we almost forget the duty of nourishing our mortal bodies," Maria Cowles wrote to her brother of her school environment. Cowles referred to students' participation in these invigorating discussions as each one "throw[ing] her mite into the treasury."[105] Students were so enthusiastic about learning that Mary Lyon was concerned they were putting study ahead of her evangelical imperative to seek spiritual salvation. In a private letter written while teaching at her school in Buckland, Lyon wrote, "The thought that some, who were beginning to think about their eternal interests may here become so much absorbed in their studies, so much interested in the business of the school, as to exclude God from their hearts, is truly painful."[106]

For many women, not being permitted to receive a full education was untenable, and they spoke passionately about their desire to learn. Almira Phelps was among the most eloquent: "To a mind thirsting for the pure waters of knowledge," she wrote, "it is tantalizing in the extreme, to be condemned to see the fountain in the far off distance, to taste a few scanty drops and yet never allowed to gain a nearer access."[107] A woman identified as S.F.W. asked plaintively, "Why must woman be denied to drink of the deep fountain of knowledge, and to experience the rich delight it infuses into the mind?" She planned to pursue education for herself, so that "*those treasures shall yet be mine*; though years of toil and self-denial be the *sacrifice*."[108] A member of Young Ladies' Association of the New-Hampton Female Seminary flatly stated that as "an individual, permitted by right, and bound by duty, to pursue her own happiness, woman should bestow the highest possible attention upon the development of her mind." Knowing that too few women were able to pursue higher education, she cried despairingly, "Oh how many a burning sentiment, how many a heavenly aspiration, has been quenched in the unfathomed mine, as diamonds destined never to imbibe the solar ray!"[109] Female students at Oberlin Institute described themselves in 1839 as scholars who "bent hither their footsteps ardently panting for knowledge & improvement."[110]

Principals of girls' schools understood their students' desires to pursue advanced subjects. Administrators offered rewards for good work. These rewards were not increased leisure time, nor a break from study, but an opportunity to pursue advanced study. Boston's High School for Women permitted third-year students to take up advanced logic, algebra, Latin or French "as tokens of merit and incitements to industry."[111] A teacher wrote in her private journal about an incident in which a student wanted to change classes. On hearing that extra work would be required, the student "looked upon me with such a cheerful, happy expression, saying, 'Well, I can do *that*; any thing that will make me learn.' "[112]

Women expected the gratification that came from study to continue into their adult lives. "I am so glad that you are so pleased with the study of history," a woman wrote to her niece away at school. "It is my favourite [*sic*] study," she went on, expressing her "hope [that] we will have many pleasant hours reading & talking about it."[113] Another woman wrote to a friend, "It is a thick snow storm, and I have been entertaining myself with Plutarch." She wished she had "something entertaining to write you, but I have no news but from the regions of the mind, there it should always be bright, and calm sunshine," confiding to her friend that "there is great happiness in cultivating the mind."[114] A young mother of an infant and two toddlers wrote that, "trifling as are my own acquisitions, I feel . . . an unspeakable gratitude that I can now and then sweeten my toils by the pleasures of reading and reflection, of imagination and composition; and I think I can form some *faint* idea of the mine of wealth and enjoyment *she* must possess, whose powers of mind have been *extensively* cultivated."[115]

Rhetoric surrounding the love of learning was strikingly similar for men and women. Women were allured to the temple of science, and thirsted to drink from the fountains of knowledge.[116] But this hungering and thirsting for knowledge was just as likely to be invoked on behalf of males. New York's high school for boys, for instance, set out in 1826 to "allure" boys to "the portals of learning," and "to secure their ardor in advancing along the gradations of the temple, until we have excited in them a genuine relish of its beauties, and a manly thirst for the treasures it contains."[117]

# "A Mysterious Connection Exists between Body and Mind": Physical Education

Educators were concerned with more than just intellectual education. Essays, addresses, and school catalogs all contained the common refrain of the need for education that devoted itself to "perfecting our whole nature, intellectual, physical and moral."[118] Educators and doctors frequently discussed physical education in the 1820s and 1830s, and they saw it as a need for males and females alike. Physical education served several purposes, according to the rhetoric of this era. First, of course, was a concern for health. A pulmonary specialist wrote in 1836 that the "period of confinement in schools is much too long for the health of all children," and urged attention "to the physical improvement of the general system."[119]

Believing that body, mind, and spirit were inseparable, educators therefore believed that exercise not only promoted physical health, in itself a worthy goal, but that physical health in turn supported both intellectual and moral development. A "mysterious connection exists between the body and mind," said Professor Mayes of Transylvania University in Kentucky, and the "judicious teacher, observing this connection, introduces into his school Gymnastic and Calisthenic exercises."[120] Furthermore, educators believed that the link between physical health and intellect and morality was as true for women as for men. "We know not why [physical education] should ever be disjoined from intellectual and moral culture," wrote an essayist on female education in 1826, given that "[h]ealth of body has in their case . . . the same influence over vigor of mind, as in that of the 'lordly sex.' "[121] The equation of bodily vigor with mental vigor appears in much of the literature. "By healthful exercise," wrote a physician, the brain's "strength is promoted, and its vigorous energies secured," while "inactivity impairs its tone" and even causes imbecility.[122] An article that reviewed recently published books on calisthenics and gymnastics noted that American girls were far less healthy and robust than European girls. "Nor need we wonder," the author concluded, "at the very small proportion of vigorous, active, useful minds" among American women.[123] The author of an article in the *Boston Medical and Surgical Journal* argued that "no man or woman can have a calm and powerful intellect, capable of effecting great things, who has not a sound body."[124] The very presence of the argument suggests that not only this anonymous author, but possibly also the editors of the medical journal and the editor of the education journal that reprinted the piece, believed that women could indeed possess calm and powerful intellects, and that the likelihood of their possessing such an intellect was dependent, as was men's, on good physical health.

Historians have emphasized the dire warnings of loss of health—particularly of reproductive health—directed at female scholars in the late nineteenth century.[125] There were similar warnings in the 1820s and 1830s, but with two essential differences. Perhaps most striking, threats that too much study would cause women to become unable to bear children were absent. Equally important, although doctors told young women that *lack of exercise* could damage their health, they did not attribute physical deterioration to study itself.[126] Exercise and study should be pursued in tandem, urged educators and doctors. Moreover, this latter argument was made about boys as well as girls. Physicians published articles containing tales of students going insane from too much attention to their books. One such anecdote involved a young man whose mental health degenerated into such a state of confusion that one day he wandered away from school, boarded a ship, and suddenly came out of his fog as he wandered through the streets of Montreal. The culprit of his mental decay was too much studying ("particularly the Latin Grammar!") while "taking very little exercise."[127] Another doctor claimed to "have met with many distressing examples of young men" who were compelled to leave college because of the failure of their health due to excessive study; these men risked having "their constitutions permanently injured," or even "the loss of life," if they did not moderate their study with exercise.[128]

Viewing too much study and too little exercise as a problem with equal consequences for men and women, not as a gender-specific problem, educators also propounded the solution. A physician who wrote for the *Journal of Health* stated his opinion that the "bodily exercises of the two sexes ought, in fact, to be the same," and that "both should be permitted, without control, to partake of the same rational means of insuring a continued flow of health."[129] Similarly, an 1826 article on gymnastics insisted that "corporeal exercise is necessary to both sexes."[130] Some schools took this position, as well. Buffalo's coeducational high school advertised in 1828 that gymnastics would be taught to both male and female students.[131]

Some of the discussion surrounding physical education for females posed surprisingly direct challenges to notions of delicate femininity. One proponent of calisthenics wrote that men ought not to allow women "the privilege which some of their own number, and certain mawkish, male sentimentalists would claim for them, of being such frail and tender beings, as to be little better than interesting invalids."[132] Another writer who thought that women's health would improve through vigorous exercise recommended "a slight infusion of the Spartan contempt of hardship."[133] In 1826, the journal *Medical Intelligencer* published a letter from William B. Fowle, principal of a monitorial school for girls in Boston. Fowle was determined that his students should pursue gymnastics, and expressed disgust for "the prevailing notions of female delicacy and propriety [that] are at variance with every attempt to render females less feeble and helpless." Looking for help in designing an appropriate exercise program, he "read all the books I could find," but found little that was useful. "It seemed," Fowle wrote, "as if the sex had been thought unworthy of any effort to improve their physical powers." After attending lectures on physical education, he installed bars and pulleys in the school, which the students eagerly learned to use. Again registering contempt for certain ideals of delicacy, Fowle admitted that "many hands have been blistered, and perhaps a little hardened by the exercises, but I have yet to learn that the perfection of female beauty consists in a soft, small, and almost useless hand, any more than in the cramped, diminutive, deformed, and useless feet of the Chinese ladies."[134]

Apparently some form of gymnastic exercise and use of weights rapidly became fairly common, although the time allotted for it varied greatly. A write-up of Catharine Beecher's Western Female Institute in Cincinnati, Ohio, praised Beecher for requiring half an hour of calisthenics each day, as opposed to what the writer saw as the more common practice of shorter periods for exercise.[135] An anonymously published book entitled *A Course of Calisthenics for Young Ladies in Schools and Families* recommended the use of four- and five-pound handheld weights, expressing disapproval of the use of eight- and nine-pound weights.[136]

As Fowle's report indicates, educators often succeeded in making the exercises appealing to students. An Alabamian at school in New York wrote home to her parents that she "had a fine time going to calisthenics" and happily reported that she was "much stronger and much more courageous" than she had been before.[137] This experience was not universal, however. The author of an article in 1827 promoted regular exercise, advocating that teachers and parents find ways to make exercise fun, and not to perpetuate the "dismal dejecting sights and sounds . . . [of] a row of boarding school girls swinging their 'dumb bells' with the rueful air of prisoners in a penitentiary."[138]

Walking was a far more popular form of exercise than calisthenics. Numerous schools noted in their catalogs that they intended that students use daily recesses for exercise, most often walking in the open air. Schools took pains to announce their healthful locales, including mentions of "sufficient ground . . . to admit of exercise and recreation within its own limits."[139] Arcade Ladies' Institute in Rhode Island boasted of its "extensive promenade beneath a glass canopy, equally protected and ventilated at all seasons," which furnished "an accommodation for recreation and exercise."[140] More often, school catalogs refrained from detailing the type of exercise students would pursue, and simply reassured parents and students that physical education would "receive all that attention which is demanded by regard to health."[141]

Women's strong desires for intellectual cultivation propelled much of the growth of women's educational opportunities. Hungry to learn, scores of women pursued education assiduously. They worked hard to enroll in institutions of higher schooling, and once enrolled, worked hard to take advantage of being there.

Although there were people in the 1820s and 1830s who believed that women were not capable of higher learning, a great many others believed that women were. Extending the Enlightenment belief that rational thought separated humans from animals, these people saw women as fully human and therefore capable of higher schooling. Educators wanted both females and males to learn intellectual independence, to think for themselves and to critically assess what they learn, and to be active learners. This emphasis on critical thinking fit in well with a republican ideology that asserted the need for an educated and informed citizenry. Given the cultural belief in the range of women's influence on both family and society, it is not surprising that many people saw the need for women as well as men to be educated and to think critically, and not to be swayed by demagogues.

Institutions calling themselves colleges continued to be predominantly male-only institutions in the 1820s and 1830s. But institutions for women increasingly offered curricula that was nearly identical to that taught in men's colleges. Some proprietors named their schools for women "collegiate institutes," and many advertised that they offered a collegiate course. A reigning paradigm among educators was that of the need for mental discipline, and that any subject that was taught well and in depth could provide the necessary mental discipline. Educators promoted the same basic curricula for both women and men, believing as most did in their equal intellectual capabilities. Educators encouraged students to perfect their whole natures, to train and improve both their minds and their bodies; female as well as male students needed exercise to sustain overall health.

Beliefs in intellectual equality did not necessarily imply that educators championed political or social equality, however. Some women, such as Lucy Stone, may have used their education as springboards for claiming political and social rights, but this was not likely the intent of the founders or instructors in most seminaries. Further, the opportunities available to some white middle-class women were not universally available. There was great promise in educational possibilities in this era, but there were also great limitations in the higher education of women.

# Chapter 6

## Possibilities and Limitations

### Education and White Middle-class Womanhood

Advanced education for women and men was more similar than it was different in both curricula and pedagogy between 1780 and 1840. Although there were those who believed that brains were sexed and that women's brains were less capable of study, the predominant rhetoric was one of equal ability. Educators believed that the type of learning that would best prepare women for their future roles was basically the same education that prepared men for theirs. There was a general consensus that women had the same intellectual capabilities as men, could enjoy intellectual pursuits as much as men, and that their lives would be similarly enriched. Teachers and students alike were motivated by a strong belief in the value of learning for its own sake, and the joys inherent in learning.

The most salient difference in educational opportunities was that of class and race, not gender. Only a small group of mostly white and middle-class people attended academies, seminaries, and colleges. By the 1820s and 1830s, that experience became part of the formation of a middle-class identity and the consolidation of middle-class cultural imperatives. Expanded higher education was an essential component of the new middle-class's self-creation. By common practice, the cost of schooling, and some-times by law, institutions of higher learning excluded the poorest parts of the population and virtually all African Americans.

As white middle-class women gained cultural capital through their education, some grew dissatisfied with their political and social subordination within their class group. Intended or not, advanced education equipped women with the confidence and skills they needed to forge a women's rights movement in the late antebellum period.

## Education and the Respectability of the Middle Class

In spite of the proliferation of institutions for higher learning, relatively few women were able to avail themselves of the opportunities that did exist. The women who attended academies, seminaries, high schools, and collegiate institutes were an elite group. Virtually all of them were white and middle or upper class. Even if these schools had been free, which none of them were, most poorer families could not survive without the earnings of teenagers or young adults who might have attended seminaries. Some students paid tuition by alternately working in factories or teaching for a few months, and then attending school for a term, but the poorest families could not manage even this.[1]

Some schools provided opportunities for girls and women from lower income families. Mount Holyoke kept costs low by instituting a system in which students and faculty performed all the domestic labor. Its founder, Mary Lyon, intended to provide high-quality education for those in the "common walks of life."[2] Oberlin College also kept costs low by having students perform manual labor.[3] In addition, some schools provided scholarships for a number of students, and literary societies worked to fund education for members of the "deserving poor."[4] Troy Female Seminary provided scholarships for poorer students who intended to become teachers, but scholarship students only comprised about ten percent of the student body.[5] While these scholarships no doubt had a huge impact on the lives of those women who received them, the majority of Troy's students were from wealthy enough families that they did not need scholarships. In fact, when Emma Willard approached the New York state legislature asking for funds for female education, she did not argue that the state should help provide education for those who otherwise could not afford it. She argued that state funding would improve the quality of education for women, and that the state had as much of a responsibility to provide funds for female as for male education.[6]

Opportunities for advanced education were particularly slim for African American women, although exact data are difficult to compile. Fifteen

African American women graduated from Oberlin College before the Civil War, and others attended without graduating.[7] The Lexington (Massachusetts) Normal School accepted African American students, both male and female, when it opened in 1839.[8] But support among whites for the education of African American females was so low that when the white Quaker teacher Prudence Crandall admitted one black student into her school in Connecticut in 1832, white parents withdrew their daughters. When Crandall then reopened the school as a school for "Young Ladies and Little Misses of Color," white townspeople harassed students and jailed Crandall.[9]

African American women's desire for education was as great as that of white women's. Certainly these women pursued learning with the same assiduity, even when their options were more limited. African American women organized literary societies in Lynn and Boston, Massachusetts, in Providence, Rhode Island, and in Rochester, Buffalo, and New York; in the 1830s there were at least three such societies in Philadelphia alone.[10] Perhaps Maria Stewart made the most poignant statement of this longing in an 1831 essay for *The Liberator* in which she asked, "How long shall the fair daughters of Africa be compelled to bury their minds and talents beneath a load of iron pots and kettles?"[11] The situation improved slightly in the North in the 1850s, when several more normal schools admitted African American women, but clearly most whites considered higher education as appropriate primarily for themselves.[12]

Seldom explicitly discussed, one of the most important functions of antebellum academies and seminaries was their role in class formation and consolidation. Schools may have offered scholarships to the genteel poor, but seldom if ever to the abjectly destitute, women of color, or immigrants. Most seminaries made clear that they were educating women to be *ladies*, and the word "ladies" was in the name of many schools.[13] That word included an implicit understanding that ladies were white and that they shared certain middle-class values.[14] The process of class formation included "an impulse toward self-definition, a need to avow publicly one's own class aspirations."[15]

Middle-class status was not solely linked to economic position. Indeed, as discussed earlier, "vicissitudes of fortune," in the form of a volatile market meant that one's economic situation could change drastically. Class status, therefore, had to be linked to something other than finances. Instead of material wealth alone, the middle class created an identity as a group that subscribed to certain sets of principles. One set of values revolved around a work ethic and a sense of personal responsibility. Industriousness, hard work, punctuality, and sobriety all figured prominently. Another set of values revolved around self-improvement, appropriate use of leisure time, and ideas about what it meant to be cultured. Education, therefore, was a major key to class status. The newly forming middle class defined itself against the urban poor, which it typed as "ignorant, careless," rife with "inebriation, squalid wretchedness, Sabbath profanation, and vices."[16] One way that the middle classes separated themselves from the "lower" classes and inculcated its own values was by education.[17]

Demonstrating that middle-class status was about more than economic condition, New England female factory workers established themselves as members of the middle class in the 1820s and 1830s. Tough economic times meant that the "factory girls" could not, at that moment, anyway, be domestic "ladies." But they could exhibit the morals and habits of their class. Visitors to factory cities commented on the workers dressing like ladies, wearing "scarves, and shawls, and green silk hoods," and carrying parasols. Visitors defined the workers against perceptions of what lower class women should look like; the "superior" factory workers were "not sallow, nor dirty, nor ragged, nor rough. They have about them no signs of . . . low culture . . . ."[18] Proponents of industrialization encouraged women to come work in the mills by representing factory workers as embodying middle-class values. When their factory shift ended, the workers played the pianos in their boarding house parlors, studied together, formed reading and self-improvement circles, and even published their original essays and poems in journals of their own making. In short, industrialists projected the image of workers living a middle-class life that included cultural activities and the pursuit of knowledge.

If factory owners capitalized on these middle-class values, proprietors of academies and seminaries certainly also vied for students on these grounds. Few schools were as blunt in their appeal to class biases as New Haven Young Ladies' Institute. Its catalogs from 1829 and 1830 stated that a benefit of a boarding school education, as opposed to private tutoring at home, was "well regulated intercourse with virtuous and intelligent associates." Boarding school life would prepare students for social life as adults. However, the proprietor asserted, "it is not to every kind of society that we would attach so great a value. . . . A school must be select," he went on, choosing only the "children of families, who are themselves refined, and who value refinement."[19]

Even without strong class assertions, many academies and seminaries trumpeted their adherence to and promotion of middle-class values of respectability, morality, industry, thrift, and order. Respectability was key to middle-class status, and referred to a person's good character and conduct; someone could be respectable in this sense even while being in material straits due to economic cycles.[20] Because respectability was so central to middle-class identity, some schools advertised that they would produce ladies who were "respected."[21] Even the students' visitors had to pass muster, as, for instance, South Carolina Female Institute would admit only "respectable female relatives" during visiting hours.[22] Alabama Female Institute promised to operate the government of the school based on appealing to students' "principles of obvious propriety and obligation, and to the better feelings of the soul" and to "discipline the social and moral feelings."[23] A speaker at a New Jersey seminary praised the "propriety" of education for women.[24] Students at Rhode Island's Arcade Ladies' Institute received an "appropriate" education that encouraged "the highest moral culture."[25] Promotional literature for Ballston Spa Female Seminary promised that an education there would "improve and refine both [a student's] social and moral nature."[26] Students at seminaries would learn to "improve their own style of manners and general deportment," and to value "personal neatness and cleanliness."[27]

Attention to manners coexisted with academic study. Lexington Female Academy, which taught such academic subjects as mathematics, Latin, and Greek, also promised that "Manners and Morals will form a prominent object of attention, throughout the whole course of studies."[28] Similarly, Roxbury Female School, which promoted "the very laborious and unfashionable task of *intense thinking,*" also gave rewards for "lady-like deportment."[29] Brooklyn Collegiate Institute for Young Ladies taught "*elegance* and *propriety* of conversation and manners" along with Livy, Horace, logic, and intellectual philosophy.[30] A writer for Young Ladies' Association of the New-Hampton Female Seminary cautioned students not to slip into "inattention to personal appearance, uncouthness of manners, and a neglect of the elegant and graceful."[31] The proprietor of one seminary assured parents that "rudeness and dissipation of manners" were "controlled."[32] Another writer for the same organization despaired over the "intemperate and profane" teachers she encountered in Kentucky and Indiana.[33]

Female education, wrote the head of a ladies' seminary in New Hampshire, should promote "industry and economy."[34] Abigail Mott, author of an 1825 treatise on female education, also believed that women must be "virtuous, industrious, and economical."[35] Boston's Mount Vernon Female Seminary sought to instill "a conscientious sense of duty" in its students.[36] "Perfect punctuality, promptness, and order, is the standard presented to every pupil on entering the Seminary," wrote the principal of Le Roy Female Seminary, while Townsend Female Seminary promised to form "habits of industry, punctuality, good order, and strict economy in the use of time."[37] An advertisement for Greenfield High School for Young Ladies noted that its students came from homes where they had "been trained to habits of industry and propriety."[38]

Rhetoric emphasized values that set the middle class apart from the wealthy as well as from the poor. As discussed in an earlier chapter, some discourse ridiculed boarding schools that turned out "fops" and "dandies." Adherents of middle-class values disparaged the frivolity and wastefulness they associated with the wealthy. "All who have consumed much time in the frivolous pursuits of what is called fashionable life, must ultimately feel," said one writer in 1831, "that these things are both debasing to the character and unsufficing [*sic*] to the heart."[39] The middle class valued moderation, a mode of behavior that they thought separated them both from the licentiousness of the poor and the wasteful consumption of the wealthy. "Regulating the passions" was a value that educators claimed students would learn through higher education.[40]

Nor was this a phenomenon that pertained only to female education. Men's schools, too, created and reflected an assumption that educated men were gentlemen, but not in the aristocratic sense.[41] Promotional literature for a boys' high school in New York promised that the school would teach everything an "educated man ought" to know, because "[s]o much . . . of our respectability . . . depends upon a knowledge of the material world, and its inhabitants, and of the arts of civilized life." The moral and republican values that the principal promised to inculcate while preparing "lads" for either college or the "counting house" were "industry, prompt obedience, and exact discipline."[42] A commencement speaker at Philadelphia Academy admonished students to remember that "your future enjoyment of life, your usefulness and respectability in society, and the formation of your respective characters" depended on how they spent their time in their youth.[43]

Higher education in the 1820s and 1830s, then, was largely a class project. Far more marked than differences between men and women were differences of class and race. The white middle classes believed in education for a wide array of reasons. As concerned citizens, they saw education as a way to improve the morality of the republic. Religious devotees saw education as a way to spread their Christian beliefs, while people concerned with status saw it as a means of ensuring upward mobility. For many people, the love of learning was an end in itself. Largely unspoken as a goal, higher education also functioned as a delineator of social classes. For all these reasons, curricula and pedagogy did not need to be starkly differentiated for women and men. Indeed, for class consolidation, similarities between men and women of the same social class were as important as differences between the middle class and those outside their class.

# The Power and Limitations of Intellectual Equality

Before the late 1840s, few women or men agitated for full political or legal rights for women, yet many asserted women's intellectual equality, their right to have a good education, and the freedom to put their education to good use. Historians often date the beginning of the women's rights movement to the Seneca Falls convention in 1848, and indeed that probably marked the first time that women and men gathered publicly to discuss the rights of women. But discourses on women's rights began long before then. As Catharine Sedgwick wrote in *Means and Ends*, an advice book for young women published in 1839, "There has been a subject much agitated of late years. . . . As you come into life and mingle in society, you will hear much talk of the '*rights of women.*' "[44] Discussion of women's intellectual abilities, and the opportunities they should have to exercise those abilities, played a prominent role in the emerging discourse on women's rights.

Thomas Gisborne and Charles Butler both opposed women's political equality. In *An Enquiry into the Duties of the Female Sex*, Thomas Gisborne took note of "some bold assertors [*sic*] of the rights of the weaker sex stigmatizing, in terms of indignant complaint, the monopolozing [*sic*] injustice of the other; . . . upholding the perfect equality of injured woman and usurping man in language so little guarded, as scarcely to permit the latter to consider the labours of the camp and of the senate as exclusively pertaining to himself." Gisborne wrote this attack on advocates of women's rights in 1796, demonstrating the existence of this debate by at least that date. Forty years later, Gisborne's language continued to resonate. Charles Butler lifted this passage verbatim (as he did much else of Gisborne's work) and included it in his book *An American Lady*, published in 1836.[45] Clearly neither Gisborne nor Butler had any patience for women who thought they could step in to men's roles.

Yet both of these opponents of women's social or political equality were strong champions of women's intellectual excellence. Gisborne urged women with "strong mental powers" to exert them.[46] Butler, for his part, roundly took to task women who "foolishly affect to be thought even more silly than they are," and who "exhibit no small satisfaction in ridiculing women of high intellectual endowments, while they exclaim, with much affected humility, . . . that 'they are thankful *they* are not geniuses.' " Butler wryly commented that, "though we are glad to hear gratitude expressed on any occasion, yet the want of sense is really no such great mercy to be thankful for."[47] To Gisborne and Butler, female intellectual abilities and attainments were laudable, yet were not a basis for political or social equality.

The realm of intellect was not the only arena in which some people urged forms of equality while stopping short of proclaiming full political and social equality. The evangelical fervor that swept much of the nation in the 1820s and 1830s included many advocates of female preaching, as well as advocates of female church members' right to vote on matters of church business. Historian Catherine Brekus documented sects that allowed women to vote in church, but also noted that "these sects never claimed that women should be allowed to vote in state [elections] as well." Further, ministers who allowed women to preach and teach still maintained a belief in female subordination to men. Some men and women agreed with the minister Ephraim Stinchfield who believed that "if a woman has a gift, she has as good a right to improve that gift as a man," but exercising that gift, even in the pulpit, did not change the essential structure of men as "the head in the affairs both of church and state, as well as in his family."[48]

The three most famous school founders, Emma Willard, Catharine Beecher, and Mary Lyon, held views that emphasized sex-based differences. For none of them, however, did their beliefs that there should be political and social distinctions between men and women lead to a belief in intellectual difference. Regardless of their religious views or their views on shaping the social or political role for women, Willard, Beecher, and Lyon clearly believed in women's high intellectual capacities and their right to have those capacities fully developed.

Emma Willard used ideologies regarding men's and women's different responsibilities in life to argue for improved education for women. Willard began teaching in a village school in Connecticut in 1804. Three years later she moved to Middlebury, Vermont, where she was struck by the differences between educational opportunities available to men in Middlebury College and what had been available to her. Here she first began formulating her desire to open "a grade of schools for women higher than any heretofore known."[49] As early as 1809 she wrote a plan for state support of women's education, which she proposed to the New York state legislature in 1818. In this plan, she argued that female education suffered from the vagaries of a market system that forced proprietors to teach only courses for which parents and students were willing to pay. State support of male education, on the other hand, allowed instructors to focus on academic excellence.[50] Willard, then, argued that men and women had equal rights to a good education and had an equal claim to state financing of that education. She based this argument on women's roles as mothers who "have the charge of the whole mass of individuals, who are to compose the succeeding generation," and who therefore needed a solid education to prepare for this work.[51]

Willard's professional life supported her belief in intellectual equality without overtly challenging middle-class women's social roles. The school that Willard established, Troy Female Seminary, "bore a remarkable resemblance to the contemporary men's colleges." The curriculum included geometry, algebra, botany, chemistry, modern languages, Latin, history, philosophy, geography, and literature.[52] She deliberately chose, however, not to call her institution a college so that it would "not create a jealousy that we mean to intrude upon the provence [sic] of man."[53] To reassure parents and the public generally that she did not intend for women to renounce their own station, students at Troy also learned embroidery and other forms of needlework.[54] Willard did not plan to disrupt the social order. She referred to men as "the only natural sovereign" of a family, and told students to "above all preserve feminine delicacy." She sat down during the public speeches she gave because she considered it unfeminine for women to address a crowd; seated, she was merely engaging in conversation.[55] One scholar of Willard stated that " 'subordination' was one of her favorite words."[56]

Yet Willard also engaged in a wide array of activities beyond motherhood and rarely subordinated herself to anyone. She not only founded and ran a high-caliber institution for women, and took it upon herself to propose legislation to the State of New York, but she also wrote a dozen highly successful texts. Her geography textbook was published in 14 printings between 1822 and 1847; her U.S. history text enjoyed 53 reprintings between 1828 and 1873, and was translated into German and Spanish; and her world history text had 24 printings between 1835 and 1882.[57] She ran for, and was elected to, the office of supervisor of schools in Kensington, Connecticut— an election in which women could not vote. In the 1840s, she traveled extensively (reportedly 8,000 miles just in 1846) promoting the common school reform movement.[58] Closer to home, Willard insisted that her fiancé sign a marriage settlement that denied him access to any of the substantial property she brought to the marriage, in addition to any money or property she might acquire during the marriage. In 1839, after less than a year of marriage, she claimed fraud and cruelty, and filed for divorce in a highly publicized case.[59] Thus, Willard embodied complicated notions of womanhood. She built a seminary largely on the argument that women needed education to prepare for their all-important work as mothers, and she promoted ideals of gentility and demure femininity. Yet she also argued that women and men were equally capable intellectually and needed the same type of education, even if they put it to different uses. She wrote and published numerous books and amassed a small fortune, from which she consciously and adroitly prevented her husband from benefiting.

Catharine Beecher, daughter of the renowned minister Lyman Beecher, widely publicized her views that women were morally superior to men, and that because of this moral superiority, their work as wives, mothers, and teachers was crucial for forming and sustaining the virtue of the nation. Beecher believed that women preserved their moral superiority in part by their removal from the temptations of the world. Confined to the domestic sphere, they could not be sullied by the iniquity of the outside world. In order to retain their moral superiority, therefore, Beecher opposed women working publicly for most reform movements, including women's rights and abolition. However, she actively promoted the expansion of women's roles and opportunities in the profession of teaching. As teachers, she urged women not only to leave their hearths for the schoolroom, but she also urged them to leave New England altogether and move to the West to help Christianize the nation. Her vision then was of women as people with a special moral calling, which they should put to use within the home and the schoolroom.[60]

At the same time, Beecher attempted to elevate women's domestic roles. She openly discussed the hard labor of housework and the essential contributions of women as the backbones of society. She also pushed for higher wages for female teachers, promoted teaching as a profession for women, and argued that women should be able to be financially self-sufficient through their labor. Furthermore, her statements regarding women's sphere being that of the domestic circle did not imply a base subservience of women to men. Instead, Beecher argued, the "true attitude to be assumed by woman, not only in the domestic but in all our social relations, is that of an intelligent, immortal being, whose interests and rights are *every way* equal in value to that of the other sex. . . . And every woman is to *claim* this, as the right which God has conferred upon her."[61] Beecher, then, believed that "woman's sphere" was in the home and the classroom, and that women should leave the realm of politics to men. At the same time that she emphasized women's morality, though, she also emphasized women's intelligence. A "True Woman," for Beecher, was not merely pious, and certainly was not abjectly servile. She was an intelligent person with a right to both education and a means to support herself.

Mary Lyon, founder of Mount Holyoke Female Seminary, refused to be hemmed in by notions of appropriate behavior for middle-class women. When she was criticized for riding from town to town in her fundraising efforts, she responded, "What do I that is wrong? I ride in the stage coach or cars without an escort. Other ladies do the same. . . . If there is no harm in doing these things once, what harm is there doing them twice, thrice, or a dozen times? My heart is sick, my soul is pained with this empty gentility, this genteel nothingness."[62] Unlike Beecher, Lyon had little interest in fitting in to higher status society.

Lyon believed that a God-ordained difference between the sexes required distinctions in some respects but not in others. She set out her position clearly in a private letter to Catharine Beecher in 1836. Beecher had begun advocating higher salaries for female teachers, urging teaching as a respectable way for women to gain social status and financial independence. Lyon disagreed. She was not interested in issues of women's equal right with men to attain financial self-sufficiency. "Let us cheerfully make all due concessions," she wrote to Beecher, "where God has designed a difference in the situation of the sexes, such as woman's retiring from public stations, being generally dependent on the other sex for pecuniary support, &c."[63] This was not particularly a comment on women's right to earn money. Lyon was not arguing about whether women had a right, relative to men's right, to earn an income. That question simply did not interest her. Lyon was deeply religious and believed that leading a Christian life and leading others to that life were the only matters of real import. Her disinterest in the salaries of teachers was not a result of her belief about female subordination, or her opposition to a movement to raise women's status. Instead, it reflected her lifelong renunciation of material comfort for *all* Christians, male or female. A supporter of the revival of the theology of Jonathan Edwards, Lyon believed in self-denial and in the "doctrine of disinterested benevolence."[64] Women's motivation to become teachers, she felt, should not be monetary but rather should be to fulfill the biblical injunction to love thy neighbor as thyself.

Lyon apparently had a narrow definition of the "public stations" from which women ought to retire, given that there were many public arenas in which she thought women belonged. She trained women to be both teachers and missionaries, both of which took women away from their firesides. Lyon herself traveled alone in her fundraising ventures for Mount Holyoke Female Seminary, wrote articles promoting the seminary, and organized a coalition of men to support her venture. She was also more than willing to work behind the scenes when this seemed most politically expedient. In a private letter to Zilpah Grant about building public support for a new seminary, Lyon wrote, "It is desirable that the plans relating to the subject should not seem to originate with *us*, but with benevolent *gentlemen*."[65] She did not say that, as women, they had a moral duty to stay behind the scenes. Like many female activists, her position was strategic, not ideological.

Although Lyon was not interested in pressing for a culture in which women could take credit publicly for their ideas, or in which women took on explicitly political roles, she also did not believe that every aspect of men's and women's lives should be different. If women were to agitate for equal treatment in any regard, Lyon hoped it would be within the contexts of education and religion. She wrote, "O that we may plead constantly for her religious privileges for equal facilities for the improvement of her talents, and for the privilege of using all her talents in doing good!"[66] She spent her life trying to provide those "equal facilities" in the institutions in which she taught.

Willard, Beecher, and Lyon, then, held some similar and some different views of what realms of activity were appropriate for women and what were not. All seemed to fit an acceptable model of middle-class womanhood, especially by virtue of their piety. All three also modeled assertiveness and independence. Historiographical debate has centered on whether these three seminary founders promoted or challenged the ideology of domesticity and whether domesticity had feminist potential.[67] Instead of furthering that debate, the ubiquity and power of an ideology of domesticity must be interrogated. Various school founders, teachers, students, and parents had different purposes, intents, and views of womanhood. There may indeed have been a "cult of true womanhood" emphasizing piety, purity, obedience, and domesticity, but simultaneously there were people who believed that a "true" woman was strong, courageous, self-sufficient, rational, assertive, and, above all, intelligent.[68]

Many historians have provided evidence of these alternative visions of womanhood in the early nineteenth century. Frances Cogan documented a definition of womanhood that she called the ideology of "real womanhood." According to Cogan, those who believed in this ideology valued self-sufficiency, intelligence, and physical fitness and health, and their beliefs were reflected in popular magazines and novels.[69] Mary Kelley concluded that most of the female characters who appeared in over two hundred novels, stories, and essays in the antebellum era had the attributes of strength, activity, and independence.[70] Laura McCall, analyzing the premier women's magazine *Godey's Lady's Book*, found that of 234 characters in stories and essays, "*not one* possessed all four features [of piety, purity, submissiveness, and domesticity] that purportedly made up the 'true woman.' " McCall's analysis further showed that only 14 characters had 3 of the 4 determinants of "true women," and 85 had none at all.[71] When McCall analyzed 104 bestselling antebellum novels, she found that over 70 percent of the authors of these bestsellers, both male and female, created female characters who were celebrated for their independence and ingenuity.[72]

Examination of the discourse surrounding advanced education for women suggests that many seminaries promoted visions of womanhood that included attributes of independence, intelligence, and strength. As described in previous chapters, writers of prescriptive literature promoted self-sufficiency for women in both the late eighteenth and early nineteenth centuries. Further, rather than emphasizing "femininity," some people in evangelical circles praised women, especially those who worked as teachers, for being "manly." For instance, in a eulogy for educator Martha Whiting, her minister said she was "distinguished for manly and Christian independence. As a Christian, she endeavored to think and act for herself."[73] Catherine Brekus found that many evangelical groups "lauded women for their 'masculine' acts of courage as well as their 'feminine' nurture of their families."[74]

Promoters of advanced education for women often cited as a benefit of education that women would be confident enough to assert themselves and take action. For instance, in a treatise called *Observations on the Importance of Female Education*, Abigail Mott told the story of a quick-thinking woman who assertively corralled the labor of neighbors to save a house from fire. "That [women] should be taught timidity," Mott moralized, "or to consider it as an accomplishment to shrink from the appearance of danger, is a great error." The qualities she listed as important for women to have were prudence, fortitude, and presence of mind, along with virtue, industriousness, and the ability to economize.[75] Seminaries advertised that they would train students to "form a general character of self-reliance," to "depend on themselves," or to "cultivate a proper reliance on her own powers."[76] Private letters as well as published essays reinforce that many people shared these views. Marcus Stephens, a Southern plantation owner, wrote to his granddaughter who was in school in North Carolina that "women have not been treated with Justice by the male sex. . . . [H]er mind is equally vigorous as his . . . and I have known several instances in private life where women have exhibited full as much courage, prudence and strong sense as any man in like circumstances."[77] A vigorous mind, courage, prudence, and strong sense may not have been the defining characteristics of "true women," but they were attributes that many held in high esteem for women.

Whatever the intentions of educators or parents, the experiences some women had in the academies, seminaries, high schools, and colleges may have created a desire for social or political equality while simultaneously providing the foundation from which to articulate those sentiments. Advocates of higher education clearly promoted a view of men and women as intellectual equals. Furthermore, many educators inspired in students a sense of certitude in their abilities. Students remembered learning self-respect and a "stubborn faith in the capacity" of women at these institutions.[78] Men, too, encouraged women to have self-respect and confidence. A male speaker at a female seminary in 1840 told students that their minds were like steam ships: an educated woman, he said, "has within herself the power . . . of progressing and keeping her course despite . . . opposing winds and waves."[79] Another male speaker concluded an address to a women's literary society by exclaiming, "To sum up the whole in one concise rule, which, I hope, you will treasure up in your memory, and practise [*sic*] in your lives, I would say—*in your intellectual and moral being, 'call no man master'.*"[80] In these ways, women were encouraged to take themselves and their ideas seriously.

The government structure in many institutions taught responsibility, and encouraged students to think for themselves rather than simply to obey rules. The structure consistently was one of "*self-government*" based on "the dictates of enlightened reason . . . [and not] mere submission to authority."[81] The "great object of all intellectual and moral culture," according to the trustees of Uxbridge Female Seminary, was "*entire self-government.*"[82]

School government must be "rational," not simply blind obedience, said a flyer for a female collegiate institute in western Pennsylvania.[83] The Roxbury Female School had its own constitution for students to adhere to, drawn up by a committee of students. The constitution included a system of appeals for students who felt they had been given a demerit unjustly. Clearly Roxbury's students were actively involved in the school government.[84] Many institutions of higher education encouraged women to rationally assess the wisdom and fairness of rules, and some created a system in which women participated fully in the establishment of the regulations by which they would live. For some women, this situation no doubt contrasted bitterly with their relationship to state and federal government, and possibly their relationships with male members of their own families, as well.

Higher education also fostered a belief in the superior morality, not only of women, but also of the white, native-born, middle class. Not surprisingly, by the late 1840s, some educated women objected to men of "inferior" classes gaining political rights that were denied to white women. The 1848 "Declaration of Sentiments" criticized men for withholding from women "rights which are given to the most ignorant and degraded men— both native and foreigners."[85] In 1854, Elizabeth Cady Stanton said in a speech, "We [white women] are moral, virtuous, and intelligent, and in all respects quite equal to the proud white man himself, and yet by your laws we are classed with idiots, lunatics, and negroes. . . . Can it be that you [in the New York state legislature] . . . would willingly build up an aristocracy that places the ignorant and vulgar above the educated and refined . . .?"[86] Although these attitudes existed by the late 1840s, because the antebellum woman's rights movement was so linked with abolition, this theme did not predominate until after the War.[87]

In addition to, unwittingly or not, nursing in women a dissatisfaction with their political and social status, higher education also gave students the tools to build a women's movement. At a minimum, higher education gave women confidence in their intellectual abilities. Many seminaries required students to write compositions, thereby honing their writing skills and their abilities to formulate coherent arguments. Many seminaries asked students to display their intellectual skills before large crowds at public examinations, and some students also read compositions at these proceedings. Hundreds of seminary students went on to organize institutions of their own.[88] In the process, they learned organizational skills, fundraising, management, budgeting, and networking. Historians have suggested that leaders of the women's movement learned their speaking and organizational skills from their prior work in the abolition movement, temperance, and missionary work, and indeed many women's leaders had labored in these fields.[89] But many women's leaders, including Elizabeth Cady Stanton, Antoinette

Brown Blackwell (the first female ordained minister in the United States), Lucy Stone, and Abby Kelley applied the skills and knowledge gained from their advanced educations in the creation of their roles as public activist women.[90] Their experiences in seminaries were as important in launching their activism for women's rights as their experience in abolitionism.

# Feminized Men and Masculinized Women: Opposition to Women's Education

This study ends in 1840, the year by which the three most famous female seminaries were established. In the following decades, women's education continued to expand. The greatest growth in seminaries occurred in the 1850s, when hundreds of such schools sprang up across the country. This decade also saw an increase in the number of women's schools willing to take on the name "college" rather than "seminary," as well as an increase in the number of coeducational colleges. In the 1850s, there were more than 45 degree-granting colleges open to women.[91]

Advanced education for women did not face much overt opposition until the late nineteenth century. One of the opening volleys was lobbed by Dr. Edward Clarke. Clarke, a Harvard Medical School professor, published *Sex in Education* in 1873, in which he argued that by studying too much, females misdirected blood from their "female apparatus" to their brains, resulting in "neuralgia, uterine disease, hysteria, and other derangements of the nervous system," along with eventual sterility.[92] Tied in with the newly emerging eugenics movement and a racist fear of the decline of the "white race," Clarke and his allies attacked women's education.

In the 1870s, too, women's rights activists more vocally turned to the issue of suffrage than they had in the antebellum era. Activists who had referred to their movement as the "woman's rights movement" in the ante-bellum era began referring to it as the "woman suffrage movement" after the Civil War.[93] Activists more often and more vociferously linked suffrage and education in this later period. In the 1820s and 1830s, before any organized women's rights movements, women and men argued for women's right to education, and educators specifically argued for increased opportunities for education, without calling for political equality. In the late 1840s and 1850s, activists began to appeal to class and racial biases. After the war, women's rights advocates more often used the education already attained by middle- and upper-class white women as an argument in support of women's right to vote. Education demonstrated women's intellectual capacity for making clear and rational decisions, and also connoted the moral center that was deemed peculiarly female and middle class. The moral sensibilities and cultural refinement, along with mental discipline, that white middle-class women acquired through education qualified them for the franchise. One faction of the woman's suffrage movement resorted to racism, classism, and nativism, warning "American women of wealth, education, virtue and refinement" about "the lower orders of Chinese, Africans, Germans and Irish" who would make laws for them.[94] Education became a symbol of white middle-class women's claim to political and social rights. This push for women's rights came at a time when white middle-class men were experiencing challenges from many fronts.

By the late nineteenth century, some historians argue, masculinity was "in crisis." While other historians criticize the hyperbole, most do agree that there was intense interest in this period in redefining manhood.[95] Economic downturns and cycles of severe depressions in the 1870s, 1880s, and 1890s meant that a man's financial status was not as secure as in an earlier generation. A growth in low-level clerical work meant that men might be less likely to move up the ladder in business.[96] Self-employment, which had been a hallmark of middle-class manhood, dropped precipitously; the percentage of middle-class men who had been self-employed declined from 67 percent in 1870 to 37 percent by 1910.[97] Meanwhile, working-class men—against whom middle-class men defined themselves—were in open revolt. In the last two decades of the nineteenth century, there were nearly 37,000 strikes involving seven million workers, and many of the strikes were violent.[98] Furthermore, the political clout of immigrant groups grew in this period, as immigrants "wrested political control from middle-class men in one city after another."[99] According to historian Gail Bederman, the "power of manhood, as the middle class understood it, encompassed the power to wield civic authority, to control strife and unrest, and to shape the future of the nation"; each of these meanings of manhood was now being threatened.[100]

White middle-class men responded to these threats by redrawing the boundaries of manhood. Millions of men joined male-only organizations, either secret or civic societies, such as the Freemasons and Oddfellows, and they signed their sons up for groups dedicated to turning boys into men, such as the Boy Scouts.[101] Athletics flourished as men devoted themselves to bodybuilding, fighting, and football; also flourishing were vigorous outdoor adventure groups comprised of men seeking the "strenuous life" advocated by Theodore Roosevelt. Even as Christians, men needed to be "muscular" in their faith.[102] In short, the lines "separating masculinity and femininity had become more sharply drawn; less permeable and elastic" than they had been in the early republic and antebellum eras.[103] It was precisely at this moment that women were making some of their greatest strides toward formal education.

From the 1870s on, the number of women attending institutions for higher education increased dramatically. In 1870, approximately 11,000 women were enrolled in seminaries or colleges. By 1880, the number had jumped to 40,000, and in the next 20 years, it more than doubled, reaching 85,000 in 1900.[104] Not only that, but the majority of these female students enrolled in coeducational institutions, where some professors and male students viewed them as "the feminine equivalent to the yellow peril in education."[105] In this context, there was a backlash against higher education for women.

The hostility toward women in coeducational colleges took several forms. Some universities, such as the University of Chicago and the University of Wisconsin, established sex-segregated classes for men and women. Stanford banned women from some liberal arts courses, and instituted a ratio restriction of at least three males to every female student enrolled in the university. The University of California established a junior college system for women. Other institutions set up "coordinate colleges" for women on separate campuses. Examples of this system of keeping women out of "men's" schools include Barnard for women who otherwise might have attended Columbia, Radcliffe at Harvard, Pembroke at Brown, Sophie Newcomb at Tulane, Jackson at Tufts, and the Women's College at the University of Rochester.[106] Admission restrictions and quotas did not disappear entirely until after passage of Title IX of the Education Amendments of 1972, which outlawed gender discrimination in educational institutions that accepted federal money.

Outright hostility to women's education reached its peak around the turn of the twentieth century. Even then, few argued that women should not be educated. Rather, the most strident objections were to men and women being educated together, in coeducational settings. In the creation of coordinate colleges, for instance, the implication was that it was acceptable for women to get a college education, as long as they did not get it in the same classrooms as men. Opponents of coeducation feared either the feminization of men or the masculinization of women, or both. They feared either that women's lesser abilities would slow men's academic progress by holding entire classes back, or else that women's superior abilities would demoralize men who could not achieve the same level of academic success. They simultaneously feared that women would come to coeducational colleges simply to seek husbands, and that coeducation would take the "mystery" out of heterosexuality and result in fewer marriages altogether.[107]

People who became hostile to advanced education during this period did so for many reasons. Suffrage activists were more vocal, and perhaps presented more of a threat of social disruption than agitators for women's rights had in the 1820s and 1830s. Suffrage activists also made an explicit link between women's education and their right to vote. Because birth rates for educated women were lower than for those who had not pursued advanced study, some people worried that white women would refuse—or be unable—to propagate the "white race." Finally, the proportion of women attending schools of higher learning before the Civil War was very small, but increased dramatically at the end of the nineteenth century. Not only might such women not reproduce, but they also might take over the formerly male institutions.

None of this pertained to earlier periods, however. The late nineteenth century hostility was a backlash specific to that period, and not indicative of what had come before. There was less opposition to women's education in the early republic and antebellum era than there was at the end of the nineteenth century. In the absence of immediate threats to the gender order and drawing on popular ideas derived from the Enlightenment, the Revolution and evangelicalism, antebellum higher education for women matured in a relatively friendly atmosphere. This is not to say that these were golden years for women's education. Women certainly did not have broad access to higher education, and that sad fact may partially explain why women's education was not dramatically opposed. Women did not face stiff antagonism until they achieved a critical mass, comprising roughly one-third of the student body of coeducational institutions in the late nineteenth and early twentieth centuries.

The academies and seminaries of the early republic and antebellum era paved the way for the women's college movement, not by proving women's intellectual capabilities, but by institutionalizing women's *right* to education and setting in motion a commitment to access to equal education for women. Before the twentieth century, only a tiny proportion of either men or women earned college degrees (even today, only 25 percent of the population has a college degree). Advanced education in the early republic and antebellum era included only a very small number of women, most of whom were white and either middle or upper class. Within that small group, however, most educators held an assumption of intellectual equality. Educators and students alike held an ideal of learning as one of life's great pleasures, a pleasure that was of equal value to women and men alike.

# Appendix: Institutions Considered in This Study, by State and Year of Data

## ALABAMA

Alabama Female Athenaeum, 1836
Alabama Female Institute, 1836–1838
Tuscaloosa Female Academy, 1832
University of Alabama, 1833

## CONNECTICUT

Clark's Seminary for Young Ladies and Gentlemen, 1840
Hartford Female Seminary, 1827, 1832
New Haven Young Ladies' Institute, 1830, 1839
Torringford Academy, 1831
Washington College, 1835
Watson's School, 1834
Wethersfield Female Seminary, 1826–1827

## GEORGIA

Georgia Female College, 1838
Sparta Female Model School, 1838

## ILLINOIS

Monticello Female Seminary, 1840

## INDIANA

South Hanover Female Seminary, 1838

## KENTUCKY

Knoxville Female Academy, 1831
Lexington Female Academy, 1821
M'Cullough's School for Young Ladies, 1839
Van Doren's Collegiate Institute for Young Ladies, 1832

## MARYLAND

St. Joseph's Academy for Young Ladies, 1832

## MASSACHUSETTS

Abbot Female Seminary, 1840
Amherst Academy, 1816, 1817, 1821–1828, 1832, 1839
Amherst Female Seminary, 1835
Bonfils' Institution for the Education of Young Ladies, 1828
Boston High School for Girls, 1826–1828, 1831
Bradford Academy, 1827, 1839–1840
Buckland Female School, 1826, 1829–1830
Byfield Seminary, 1821
Charlestown High School for Young Ladies, 1834
Day's Academy for Young Gentlemen and Seminary for Young Ladies, 1834
Female Classical Seminary, 1826–1827
Gothic Seminary, 1840
Greenfield High School for Young Ladies, 1830, 1836–1837, 1840
Ipswich Female Academy, 1829, 1831, 1833, 1835–1839
Mount Holyoke Female Seminary, 1835–1840
Mount Vernon Female School, 1831, 1836
Roxbury Female School, 1828, 1830
Sanderson Academy, 1817, 1822, 1828–1829
Thayer's School, 1826
Townsend Female Seminary, 1839
Uxbridge Female Seminary, 1833, 1837–1840
Wesleyan Academy, 1833
Westfield Academy, 1840–1841
Worcester High Schools (two) for Girls, 1833

# MISSISSIPPI

Elizabeth Female Academy, 1828

# NEW HAMPSHIRE

Adams Female Academy, 1824–1826, 1831
Atkinson Academy, 1815–1840
New-Hampton Female Seminary, 1834–1840
Young Ladies' Seminary in Keene, 1832–1833

# NEW JERSEY

Newark Institute for Young Ladies, 1826
Rutgers Female Institute, 1839
Spring-Villa Female Seminary, 1839
St. Mary's Hall, 1837

# NEW YORK

Albany Female Academy, 1834, 1835
Albany Female Seminary, 1827–1840
Alfred Academy, 1836–1840
Amenia Seminary, 1841
Ballston Spa Female Seminary, 1824
Brooklyn Collegiate Institute for Young Ladies, 1830
Buffalo High School, 1828
Crane's School, 1831
Geneva Female Seminary, 1839–1840
LeRoy Female Seminary, 1839–1840
Newburgh Female Seminary, 1837
New York High School for Boys, 1829
New York High School for Girls, 1826
Ontario Female Seminary, 1828, 1839
Mrs. Plumb's School, 1829
Rochester Female Seminary, 1834
Utica Female Academy, 1840

# NORTH CAROLINA

Anson Male and Female Academy, 1820
Miss Ballantine's Seminary, 1825

Charlotte Academy, 1825, 1826
Clinton Female Seminary, 1837
Mrs. Edmonds' Boarding School, 1820?
Franklin Academy for Boys, 1813
Greensboro Academy, 1821
Gregory's Boarding School, 1808
Hillsborough Academy, 1801, 1803, 1825, 1839
Hillsborough Female Seminary, 1825–1827, 1829–1830, 1838–1839
Hyco Academy for Boys, 1834
Jamestown Female Academy, 1819
Kelvin School for Young Ladies, 1828, 1831, 1835–1838
Lincolnton Female Academy, 1821–1830
Milton Female Academy, 1819
Morganton Female Academy, 1824
New Bern Academy, 1823
North Carolina Female Academy, 1823, 1826
Northampton Academy, 1835, 1837–1838
Oxford Female Seminary, 1822, 1827, 1839
Oxford Male Academy, 1839
Phillips' Female School, 1836, 1838
Pittsboro Female Academy, 1838–1839
Miss Prendergast's School, 1818
Raleigh Academy, 1811, 1835
Salem Female Boarding School, 1807, 1840
Scotland Neck Female Seminary, 1837
Simpson's School, 1839
Southern Female Classical Seminary, 1830–1834
Vine Hill Academy, 1812, 1837
Walker's Male and Female Academy, 1830
Warrenton Female Academy, 1807, 1809, 1817–1818
Williamsborough Academy, 1825
Williamsborough Female Academy, 1826–1831, 1834, 1838
Woods' Female Academy, 1839

## OHIO

Granville Female Academy, 1836
Granville Female Seminary, 1836, 1838
Oberlin Collegiate Institute, 1834–1840
Steubenville Female Seminary, 1838–1840
Western Female Institute, 1833

# PENNSYLVANIA

Andrews and Jones Select Female Seminary, 1825
Brown's Young Ladies' Academy of Philadelphia, 1787, 1795
Germantown Academy, 1832
Mrs. Hugh's Boarding & Day School, 1821
Poor's Young Ladies' Academy of Philadelphia, 1792?–1794
Rural Seminary, 1834
Susan M. Price's School for Girls, 1826
Western Collegiate Institute for Young Ladies, 1837
William Russel's School for Young Ladies, 1834

# RHODE ISLAND

Arcade Ladies' Institute, 1835
Providence High School, 1828
Smithville Academy, 1838

# SOUTH CAROLINA

Charleston College, 1824
South Carolina Female Institute, 1828

# TENNESSEE

Columbia Female Institute, 1838

# VIRGINIA

Female Academy at Sturgeonville, 1828
Mrs. Garnett's, 1825
Winchester Academical Institute for Young Ladies, 1835

# Notes

## 1  Introduction

1. Catharine Beecher, *Essay on Slavery and Abolitionism with Reference to the Duty of American Females* (Philadelphia: Henry Perkins, 1837), 98–99.
2. Catharine Beecher, "Female Education," *American Journal of Education* II (April and May 1827), 220, 484, 739; Thomas Woody, *A History of Women's Education in the United States*, I (New York: The Science Press, 1929), 320.
3. For a notable example of an advocate of advanced education for women, see Judith Sargent Murray, who saw education as an inroad for establishing "the female right to . . . equality with their brethren"; Judith Sargent Murray, *The Gleaner* [1798]. Nina Baym, ed. (Schenectady, NY: Union College Press, 1992), 707, 709. For examples of antagonists to female education due to fears of usurpation of male roles, see William Alexander, *The History of Women* (Philadelphia: J. H. Dobelbower, 1796), 66, in which he writes, "we should perhaps grudge [women] the laurels of [l]iterary fame, as much as we do the breeches." See also, "Characteristic Differences of Male and Female of Human Species," *New York Magazine* (June 1790), 337; "On Female Authorship," *Lady's Magazine* (January 1793), 72; and "An Address to the Ladies," *American Magazine* (March 1788), 244. Historian Lynn Gordon notes that in the antebellum era, most advocates of women's higher education did not support women's rights; however, people who worked for a broad women's rights agenda included the issue of access to higher education in their cause. See Lynn D. Gordon, *Gender and Higher Education in the Progressive Era* (New Haven, CT and London: Yale University Press, 1990), 19.
4. For histories of this period, see Harry L. Watson, *Liberty and Power: The Politics of Jacksonian America* (New York: Hill and Wang, 1990); Henry F. May, *The Enlightenment in America* (New York: Oxford University Press, 1976); Carl F. Kaestle, *Pillars of the Republic: Common Schools and American Society, 1780–1860* (New York: Hill and Wang, 1983); Robert E. Shalhope, "Toward a Republican Synthesis: The Emergence of an Understanding of Republicanism in American Historiography," *William and Mary Quarterly* 3rd ser., 29 (January 1972), 49–80; Louis B. Wright and Elaine W. Fowler, *Life in the New Nation, 1787–1860* (New York: Capricorn Books, 1974).

5. The classic work on Beecher is Kathryn Kish Sklar, *Catharine Beecher: A Study in American Domesticity* (New York: W. W. Norton & Co., 1976).

6. For instance, see Anne Firor Scott, "The Ever-Widening Circle: The Diffusion of Feminist Values from the Troy Female Seminary, 1822–1872," *History of Education Quarterly* 19 (Spring 1979), 3–24; Sklar, *Catharine Beecher*.

7. Woody, *A History of Women's Education*, I, especially 216ff.

8. Mary Borer, *Willingly to School: A History of Women's Education* (Guildford: Lutterworth Press, 1976); Ruth Perry, "Mary Astell's Response to the Enlightenment," *Women and History* 9 (Spring 1984): 13–40; Robert B. Shoemaker, *Gender in English Society, 1650–1850: The Emergence of Separate Spheres?* (London and New York: Longman, 1998).

9. Sarah Fatherly, "Gentlewomen and Learned Ladies: Gender and the Creation of an Urban Elite in Colonial Philadelphia," Ph.D. Dissertation, University of Wisconsin-Madison, 2000, ch. 3.

10. *Pennsylvania Gazette*, January 29, 1751; August 2, 1744; May 2, 1751; September 14, 1752; March 26, 1754; and December 27, 1759.

11. Lawrence A. Cremin, *American Education: The National Experience, 1783–1876* (New York: Harper & Row, 1980), 103.

12. Lorraine Smith Pangle and Thomas L. Pangle, *The Learning of Liberty* (Lawrence: University Press of Kansas, 1993), 1, 5.

13. Linda Eisenmann, "Reconsidering a Classic: Assessing the History of Women's Higher Education a Dozen Years after Barbara Solomon," *Harvard Educational Review* 67 (Winter 1997), 697.

14. Robert L. Church, *Education in the United States: An Interpretive History* (New York: The Free Press, 1976), 23.

15. Helen Lefkowitz Horowitz noted the more frequent use of the term "seminary" rather than "academy" in the antebellum era, and suggested that the new word "connoted a certain seriousness." According to Horowitz, school founders chose the word "seminary" to reflect their intention to train women to be teachers just as male seminaries trained men to be ministers. Helen Lefkowitz Horowitz, *Alma Mater: Design and Experience in Women's Colleges from their Nineteenth-Century Beginnings to the 1930s* (New York: Alfred A. Knopf, 1984), 11.

16. "Female High-School of Boston," *American Journal of Education* I (January 1826), 61; "Boston High School for Girls," *American Journal of Education* I (July 1826), 380.

17. *General View of the Plan of Education Pursued at the Adams Female Academy* (Exeter, NH: Nathaniel S. Adams, printer, 1831), 3.

18. *Arcade Ladies' Institute, Providence R.I.* (Providence, RI: H. H. Brown, 1834?), 7–10.

19. John Ludlow, *An Address Delivered at the Opening of the New Female Academy in Albany, May 12, 1834* (Albany, NY: Packard and Van Benthuysen, 1834), 7. William J. Reese makes note of an 1831 essay that "published an impressive review of 'Academies, High Schools, Gymnasia,' without distinguishing one category from another." See Reese, *The Origins of the American High School* (New Haven, CT and London: Yale University Press, 1995), 34.

20. Kenneth H. Wheeler, "Why the Early Coeducational College was Primarily a Midwestern Phenomenon," Paper Presented at the History of Education Society Annual Meeting, October 30, 1998.

21. Church, *Education in the United States*, 37–38.

22. The following citations represent works primarily from the 1970s and 1980s. More recent work on colleges, seminaries, and academies are cited later in this chapter. For work on the women's college movement, see Horowitz, *Alma Mater*; Patricia A. Palmieri, "Here Was Fellowship: A Social Portrait of Academic Women at Wellesley College, 1880–1920," *History of Education Quarterly* 23 (Summer 1983), 195–214; Rosalind Rosenberg, *Beyond Separate Spheres: Intellectual Roots of Modern Feminism* (New Haven, CT: Yale University Press, 1982); Margaret W. Rossiter, *Women Scientists in America: Struggles and Strategies to 1940* (Baltimore: The Johns Hopkins University Press, 1982); Patricia A. Palmieri, "Patterns of Achievement of Single Academic Women at Wellesley College, 1880–1920," *Frontiers* 5 (Spring 1980), 63–67; Joyce Antler, " 'After College, What?': New Graduates and the Family Claim," *American Quarterly* 32 (Fall 1980), 409–434; Mary J. Oates and Susan Williamson, "Women's Colleges and Women Achievers," *Signs* 3 (Summer 1978), 795–806; Patricia Albjerg Graham, "Expansion and Exclusion: A History of Women in American Higher Education," *Signs* 3 (Summer 1978), 759–773; Roberta Frankfurt, *Collegiate Women: Domesticity and Career in Turn-of-the-Century America* (New York: New York University Press, 1977); Elaine Kendall, *Peculiar Institutions: An Informal History of the Seven Sister Colleges* (New York: Putnam's, 1975); Sarah H. Gordon, "Smith College Students: The First Ten Classes, 1879–1888," *History of Education Quarterly* 15 (Summer 1975), 147–167; Jill K. Conway, "Perspectives on the History of Women's Education in the United States," *History of Education Quarterly* 14 (Spring 1974), 1–12; Mabel Newcomer, *A Century of Higher Education for American Women* (New York: Harper & Row, 1959); Marian Churchill White, *A History of Barnard College* (New York: Columbia University Press, 1954).

For work on antebellum seminaries, see Scott, "The Ever-Widening Circle"; David F. Allmendinger, Jr., "Mount Holyoke Students Encounter the Need for Life Planning, 1837–1850," *History of Education Quarterly* 19 (Spring 1979), 27–46; Elizabeth Alden Green, *Mary Lyon and Mount Holyoke: Opening the Gates* (Hanover, NH: University Press of New England, 1979); Kathryn Kish Sklar, "The Founding of Mount Holyoke College," in *Women of America: A History*, Carol Ruth Berkin and Mary Beth Norton, eds. (Boston: Houghton Mifflin Co., 1979), 177–201; Deborah Jean Warner, "Science Education for Women in Antebellum America," *Isis* 69 (March 1978), 58–67.

For overviews, see Gordon, *Gender and Higher Education*; Barbara M. Solomon, *In the Company of Educated Women: A History of Women and Higher Education in America* (New Haven, CT and London: Yale University Press, 1985); Woody, *A History of Women's Education*. On women's literacy and education, see Kenneth A. Lockridge, *Literacy in Colonial New England: An Enquiry into the Social Context of Literacy in the Early Modern West* (New York: W.W. Norton & Co., 1974); Lee Soltow and Edward Stevens, *The Rise of*

*Literacy and the Common School in the United States: A Socioeconomic Analysis to 1870* (Chicago: University of Chicago Press, 1981). On women's education in the early national period, see Linda K. Kerber, *Women of the Republic: Intellect & Ideology in Revolutionary America* (Chapel Hill: University of North Carolina Press, 1980), ch. 7. On antebellum women's education, see Scott, "The Ever-Widening Circle"; Sklar, *Catharine Beecher*; Nancy Cott, *The Bonds of Womanhood: "Women's Sphere" in New England, 1780–1835* (New Haven, CT and London: Yale University Press, 1977).

23. See, for instance, Willystine Goodsell, *The Education of Women: Its Social Background and Its Problems* (New York: Macmillan Co., 1923), 10, 17; Eleanor Flexner, *Century of Struggle: The Woman's Rights Movement in the United States* (Cambridge: The Belknap Press of Harvard University Press, 1959), 23–40; Gerda Lerner, *The Woman in American History* (Menlo Park: Addison-Wesley Publishing Co., 1971), 40; Sklar, *Catharine Beecher*, 17–19; Keith E. Melder, *Beginnings of Sisterhood: The American Woman's Rights Movement, 1800–1850* (New York: Schocken Books, 1977), 84.

24. For instance, see Solomon, *In the Company of Educated Women*, 22–23; Gordon, *Gender and Higher Education*, 16.

25. See, for instance, Kerber, *Women of the Republic*; Solomon, *In the Company of Educated Women* (especially ch. 1); Ann Douglas, *The Feminization of American Culture* (New York: Doubleday, 1977), 58.

26. Florence Howe, "Myths of Coeducation," in *Myths of Coeducation—Selected Essays, 1964—1983* (Bloomington: Indiana University Press, 1984), 210. See also, Linda K. Kerber, "The Republican Mother: Women and the Enlightenment—an American Perspective," *American Quarterly* 28 (Summer 1976), 187–205; Linda K. Kerber, "Daughters of Columbia: Educating Women for the Republic, 1787–1805," in *The Hofstader Aegis: A Memorial*, Stanley Elkins and Eric McKitrick, eds. (New York: Alfred A. Knopf, 1974), 36–59; Glenda Riley, "Origins of the Argument for Improved Female Education," *History of Education Quarterly* 9 (Winter 1969), 455–470. See also Nancy Cott, *The Bonds of Womanhood*, ch. 3.

27. Londa Schiebinger, *The Mind Has No Sex? Women in the Origins of Modern Science* (Cambridge and London: Harvard University Press, 1989).

28. Mary Kelley, "Reading Women/Women Reading: The Making of Learned Women in Antebellum America," *Journal of American History* (September 1996), 403.

29. Ronald J. Zboray and Mary Saracino Zboray, " 'Months of Mondays': Women's Reading Diaries and the Everyday Transcendental," Paper Presented at the Eleventh Berkshire Conference, June 6, 1999; Ronald J. Zboray, *A Fictive People: Antebellum Economic Development and the American Reading Public* (New York: Oxford University Press, 1993).

30. Linda Kerber contends that "separate spheres" has always been a figure of speech that has been interpreted too literally. Kerber suggests that when Tocqueville reported two distinct realms of action for men and women, he was actually reporting only on the *discourse* of separate spheres, and not on actual practice. That the discourse was so prevalent in his day perhaps indicates the

renegotiation of gender relations. The fervency with which advocates of domesticity sought to relegate women to their homes was directly proportional to the rate at which women refused to accept the standards imposed on them. See Kerber, "Beyond Roles, Beyond Spheres: Thinking about Gender in the Early Republic," *William and Mary Quarterly* 3rd ser., 46 (July 1989), 565–585.

31. Mary P. Ryan, *Women in Public: Between Banners and Ballots, 1825–1880* (Baltimore and London: The Johns Hopkins University Press, 1990).

32. Also see Jane H. Pease and William H. Pease, *Ladies, Women & Wenches: Choice and Constraint in Antebellum Charleston and Boston* (Chapel Hill: University of North Carolina Press, 1990). These authors criticize reliance on the ideology of separate spheres to explain and understand nineteenth-century women's lives. Pease and Pease argue that the complete dependence on conceptual models of duality—male/female, public/private, wealthy/poor—has obscured much of the complexity of women's lives. Instead, they argue, women's lives should be seen as drawing upon a range of choices, distinguished from each other by varying economic and social forces and regional differences. Women's lives can be seen as on a continuum from extreme dependence to extreme autonomy; few women lived at either end of the spectrum.

33. The concept of separate spheres also depends on clear demarcations between male and female, which historians and philosophers also have challenged. See, for instance, Thomas Laqueur, *Making Sex: Body and Gender from the Greeks to Freud* (Cambridge, MA and London: Harvard University Press, 1992); Michel Foucault, *The History of Sexuality*, vol. 1 (New York: Vintage Books, 1978); Judith Butler, *Gender Trouble: Feminism and the Subversion of Identity* (New York: Routledge, 1990).

34. Susan Zaeske, *Signatures of Citizenship: Petitioning, Antislavery, & Women's Political Identity* (Chapel Hill and London: The University of North Carolina Press, 2003).

35. Carol Lasser, "Beyond Separate Spheres: The Power of Public Opinion," *Journal of the Early Republic* 21 (Spring 2001), 119; Julie Roy Jeffrey, *The Great Silent Army of Abolitionism: Ordinary Women in the Antislavery Movement* (Chapel Hill and London: The University of North Carolina Press, 1998).

36. Mary Kelley, "Beyond the Boundaries," *Journal of the Early Republic* 21 (Spring 2001), 78; Julie Roy Jeffrey, "Permeable Boundaries: Abolitionist Women and Separate Spheres," *Journal of the Early Republic* 21 (Spring 2001), 79; Cathy N. Davidson, "No More Separate Spheres!" *American Literature* 70 (September 1998), 443–463.

37. See Karen Hansen, *A Very Social Time: Crafting Community in Antebellum New England* (Berkeley: University of California Press, 1994); Elizabeth R. Varon, *We Mean to be Counted: White Women & Politics in Antebellum Virginia* (Chapel Hill and London: University of North Carolina Press, 1998), 1–9; Glenna Matthews, *The Rise of Public Woman: Woman's Power and Woman's Place in the United States, 1630–1970* (New York: Oxford University Press, 1992), 3–10; Ryan, *Women in Public*, 3–18; Ruth H. Bloch, *Gender and Morality in Anglo-American Culture, 1650–1800* (Berkeley, Los Angeles, and London: University of California Press, 2003), ch. 8; Jurgen Habermas, *The Structural Transformation of the Public*

*Sphere: An Inquiry into a Category of Bourgeois Society* [1962], trans. Thomas Burger (Cambridge: MIT Press, 1994), 43–67; Lasser, "Beyond Separate Spheres," 117–120.

38. David Allmendinger's study of Mount Holyoke concludes that the seminary was essentially conservative, preserving traditional life patterns for women. Nancy Cott argues that education for women was designed to maintain gender distinctions and that seminaries only unintentionally unleashed a proto-feminism because some educated women looked beyond domestic duties. Keith Melder likewise contends that, rather than being progressive in offering women new opportunities, female seminaries perpetuated women's subordinate status. Kathryn M. Kerns asserts that Willard, Lyon, Beecher, and others viewed female seminaries as places to train women to be wives, mothers, and school-teachers. Others contest this view. Lisa Natale Drakeman places Mary Lyon at odds with social conventions, saying that Lyon did not support female passivity, while James Conforti insists that Lyon actively discouraged women from seeing the domestic circle as their primary calling in life. David F. Allmendinger, Jr., "Mount Holyoke Students Encounter the Need for Life Planning"; Cott, *The Bonds of Womanhood*, 125; Keith Melder, "Masks of Oppression: The Female Seminary Movement in the United States," in *History of Women in the United States: Historical Articles on Women's Lives and Activities*, vol. 12, Nancy F. Cott, ed. (Munich, New Providence, London, and Paris: K. G. Saur, 1993), 25–44; Kathryn M. Kerns, "Farmers' Daughters: The Education of Women at Alfred Academy and University Before the Civil War," *History of Higher Education Annual* 6 (1986), 10–28; Lisa Natale Drakeman, "Seminary Sisters: Mount Holyoke's First Students, 1837–1849," Ph.D. Dissertation, Princeton University, 1988; Joseph Conforti, "Mary Lyon, the Founding of Mount Holyoke College, and the Cultural Revival of Jonathan Edwards," *Religion and American Culture: A Journal of Interpretation* 3 (Winter 1993), 79.

39. Nancy A. Hewitt, "Taking the True Woman Hostage," *Journal of Women's History* 14 (Spring 2002), 159.

40. Thanks to Bill Reese for this analogy.

41. A useful model might be drawn from a Foucauldian analysis of power, one that suggests that there are multiple, decentered publics that exist in a variety of social and cultural places. Rather than a single public sphere, contrasted with a single private sphere, a public realm might be composed of a range of institutions, each situated within its own localized set of beliefs and values, and each operating with "its own cultural assumptions, values, and principles of inclusion and exclusion." Educational institutions are prime examples of decentered public spaces, within which educational theorists, school founders, teachers, and students constructed new sites that drew from a range of cultural assumptions regarding not only gender, but also regarding the value of education. See Iris Young, *Justice and the Politics of Difference* (Princeton, NJ: Princeton University Press, 1990), and Nancy Fraser, *Unruly Practices: Power, Discourse, and Gender in Contemporary Social Theory* (Minneapolis: University of Minnesota Press, 1989). The quote is from Hansen, *A Very Social Time*, 10.

## 2 "IS NOT WOMAN A HUMAN BEING?"

1. Anne Frances Randall, *A Letter to the Women of England, on the Injustice of Mental Subordination. With Anecdotes* (London: T. N. Longman & O. Rees, 1799), 8, 84.

2. Russel Nye, for instance, makes the point that European Enlightenment ideas took 50–100 years to cross the Atlantic, and that, therefore, the late eighteenth century is when Enlightenment ideas had their greatest impact in America. Russel Blaine Nye, *The Cultural Life of the New Nation, 1776–1830* (New York and Evanston: Harper & Row, 1960), 5.

3. Harry L.Watson, *Liberty and Power: The Politics of Jacksonian America* (New York: Hill and Wang, 1990), ch. 1.

4. Rev. Doctor Sproat, "An Address," in *The Rise and Progress of the Young Ladies' Academy of Philadelphia: Containing an Account of a Number of Public Examinations & Commencements; The Charter and Bye-Laws; Likewise, A Number of Orations delivered By the Young Ladies, And several by the Trustees of said Institution* [1789] (Philadelphia: Stewart and Cochran, 1794), 24. See also Nye, *Cultural Life*, 156.

5. "Curious Dissertation on the Valuable Advantages of a Liberal Education," *New Jersey Magazine* ( January 1787), 50.

6. John Millar, "The Origins of the Distinction of Ranks; or, An Inquiry into the Circumstances which give Rise to Influence and Authority in the Different Members of Society," [1771], in *John Millar of Glasgow, 1735–1801*, William C. Lehmann, ed. (London and New York: Cambridge University Press, 1960), 193.

7. Lord Henry Home Kames, *Six Sketches on the History of Man* [1774] (Philadelphia: n.p. 1776), 220.

8. Rosemarie Zagarri, "Morals, Manners, and the Republican Mother," *American Quarterly* 44 ( June 1992), 200. See also Linda K. Kerber, *Women of the Republic: Intellect & Ideology in Revolutionary America* (Chapel Hill: University of North Carolina Press, 1980), 26–27. For other accounts of the four-stage theory, see George Gregory, "Miscellaneous Observations on the History of the Female Sex," *Essays Historical and Moral* (London: J. Johnson, 1785), 148ff.; Lord Kames, *Six Sketches*.

9. Carl Kaestle, *Pillars of the Republic: Common Schools and American Society, 1780–1860* (New York: Hill and Wang, 1983), ch. 1; Robert E. Potter, *The Stream of American Education* (New York: American Book Company, 1967), ch. 4.

10. *The Rise and Progress of the Young Ladies' Academy of Philadelphia*, 21–22.

11. Ibid., 68.

12. For his impact on education in the United States, see Kaestle, *Pillars of the Republic*, 32–33, 87. For his views on female education, see Londa Schiebinger, *The Mind Has No Sex? Women in the Origins of Modern Science* (Cambridge and London: Harvard University Press, 1989), 172–173.

13. John Locke, "Thoughts on Education," in *The Educational Writings*, James Axtell, ed. (Cambridge: Cambridge University Press, 1968), 117, 344–346.

14. Schiebinger, *The Mind Has No Sex?*, ch. 8; Robert B. Shoemaker, *Gender in English Society, 1650–1850: The Emergence of Separate Spheres?* (London and New York: Longman, 1998), ch. 2.

15. Ibid.

16. "Tis Education Forms the Female Mind," *Massachusetts Magazine* (February 1793), 92–93.

17. "Characteristic Differences of the Male and Female of the Human Species," *New York Magazine* (June 1790), 336–338; "On the Domestic Character of Women," *New York Magazine* (November 1796), 600–602.

18. William Alexander, *The History of Women* (Philadelphia: J. H. Dobelbower, 1796), 66.

19. Hannah More, "Essays," reprinted in *The Ladies' Companion* (Worcester: The Spy Office, 1824 [orig. pub. 1792]), 51–52.

20. Thomas Gisbourne, *An Enquiry into the Duties of the Female Sex* (London: T. Cadell and W. Davies, 1798), 107.

21. "Observer, No. I.," *The Key* (January 27, 1798), 17.

22. "Thoughts on Women," *Lady's Magazine* (1792), 112, 113.

23. James A. Neal, *An Essay on the Education and Genius of the Female Sex* (Philadelphia: Jacob Johnson, 1795).

24. "Reflections on Women, and on the Advantages which they would derive from the Cultivation of Letters," *New York Magazine* (February 1790), 89. For another example, see Judith Sargent Murray, Essays LXXXVIII, LXXXIX, XC, and XCI, *The Gleaner* [1798]. In Nina Baym, ed. (Schenectady, NY: Union College Press, 1992), 702–731.

25. "Tis Education Forms the Female Mind," 93.

26. Elizabeth Hamilton, *Letters on Education* (Dublin: H. Colbert, 1801), 2–3; emphasis in original.

27. Abigail Adams to John Thaxter, February 15, 1778, quoted in Pauline Schloesser, *The Fair Sex: White Women and Racial Patriarchy in the Early American Republic* (New York and London: New York University Press, 2002), 130.

28. Cynthia A. Kierner, *Beyond the Household: Women's Place in the Early South, 1700–1835* (Ithaca, NY and London: Cornell University Press, 1998), 119; Elizabeth Steele to Ephraim Steele, May 15, July 30, and October 17, 1778, and April 29, July 13, and October 29, 1780, quoted in Kierner, *Beyond the Household*, 113; Caroline Gilman, ed., *Letters of Eliza Wilkinson, during the Invasion and Possession of Charleston, S.C., by the British in the Revolutionary War* [1839] (New York: New York Times and Arno Press, 1969), 60–61.

29. Judith Sargent Murray, *The Gleaner*, 707, 709; emphasis in original. Murray wrote poems and essays as early as the 1770s. From 1792 to 1794, she wrote a monthly column for the *Massachusetts Magazine* in which she championed the cause of women's education, among other issues. These essays were collected and published in three volumes in 1798.

30. Priscilla Mason, "The Salutatory Oration," May 1793, in *The Rise and Progress of the Young Ladies' Academy of Philadelphia*, 92–93; emphasis in original.

31. For examples, see "On the Supposed Superiority of the Masculine Understanding," *Universal Asylum* (July 1791), 9–11; "On Female Education,"

*New York Magazine* (September 1794), 569–570; "Improvements Suggested in Female Education," *New York Magazine* (August 1797), 405–408; "Outline of a Plan of Instruction for the Young of Both Sexes, particularly Females, submitted to the Reflection of the Intelligent and the Candid," *Weekly Magazine of Original Essays, Fugitive Pieces, and Interesting Intelligence* (August 4, 1798), 12–15 (August 11, 1798), 37–41.

32. "On Female Education," 569–570.

33. George Gregory, "Miscellaneous Observations," 145, 146.

34. Watson, *Liberty and Power*, 7.

35. There are many accounts of the Whiskey Rebellion. See, for instance, Louis B. Wright and Elaine W. Fowler, *Life in the New Nation, 1787–1860* (New York: Capricorn Books, 1974), 42–44.

36. For discussions of this topic, see Paula Baker, "The Domestication of Politics: Women and American Political Society, 1780–1920," *American Historical Review* 89 (June 1984), 623–624; Robert E. Shalhope, "Toward a Republican Synthesis: The Emergence of an Understanding of Republicanism in American Historiography," *William and Mary Quarterly* 3rd ser., 29 (January 1972), 49–80; Ruth H. Bloch, "The Gendered Meanings of Virtue in Revolutionary America," *Signs: Journal of Women in Culture and Society* 13 (Autumn 1987), 37–58.

37. Lawrence Cremin, *American Education: The National Experience, 1783–1876* (New York: Harper & Row, 1980), 192–200.

38. Henry F. May, *The Enlightenment in America* (New York: Oxford University Press, 1976), 193; Frederick Rudolph, ed., *Essays on Education in the Early Republic* (Cambridge: Belknap Press of Harvard University Press, 1965), 41.

39. Noah Webster, "On the Education of Youth in America" [1790], in *Essays on Education in the Early Republic*, Rudolph, ed. (Cambridge: Belknap Press of Harvard University Press, 1965), 45.

40. Ibid., 65; emphasis in original.

41. Ibid., 67.

42. See Ruth H. Bloch, "The Gendered Meanings of Virtue in Revolutionary America"; Jan Lewis, "The Republican Wife: Virtue and Seduction in the Early Republic," *William and Mary Quarterly* 3rd ser., XLIV (October 1987), 689–721; Rosemarie Zagarri, "Morals, Manners, and the Republican Mother."

43. "Female Influence," *New York Magazine* (May 1795), 297–304; "An Address to the Ladies," *American Magazine* (March 1788), 446. See also, John Swanwick, *Thoughts on Education, Addressed to the Visitors of the Young Ladies' Academy in Philadelphia, October 31, 1787* (Philadelphia: Thomas Dobson, 1787); "The Influence of the Female Sex on the Enjoyments of Social Life," *Universal Asylum* (March 1790), 153–154.

44. Ibid., 69.

45. See Daniel Scott Smith, "Parental Power and Marriage Patterns: An Analysis of Historical Trends in Hingham, Massachusetts," *Journal of Marriage and the Family* 35 (August 1973), 419–428; Daniel Blake Smith, *Inside the Great House: Planter Family Life in Eighteenth-Century Chesapeake Society* (Ithaca, NY: Cornell University Press, 1980); Jan Lewis, *The Pursuit of Happiness: Family and Values in Jefferson's Virginia* (New York: Cambridge University Press, 1983).

46. Scholars are in some disagreement about the extent of the companionate ideal in the South. For a discussion of this, see Anya Jabour, *Marriage in the Early Republic: Elizabeth and William Wirt and the Companionate Ideal* (Baltimore and London: The Johns Hopkins University Press, 1998), 3–4, 171–172, fn. 4.
47. Lewis, "Republican Wife," 699. See also Linda K. Kerber et al., "Beyond Roles, Beyond Spheres: Thinking about Gender in the Early Republic," *William and Mary Quarterly* 44 (July 1989), 577; Mary Beth Norton, *Liberty's Daughters: The Revolutionary Experience of American Women, 1750–1800* (New York: HarperCollins, 1980), ch. 8.
48. "On Marriage," *New York Magazine* (December 1796), 656–657. See also, "Instructions preparatory to the Married State," *New York Magazine* (July 1797), 374, where women are urged, in choosing a husband, to look for "a comfortable companion for life."
49. For further discussion of this ideal, see Jabour, *Marriage in the Early Republic*, especially chs. 1 and 2.
50. "On Matrimonial Obedience," *Lady's Magazine* (June 1792), 64, 66.
51. "On the Education and Studies of Women," *New York Magazine* (October 1797), 543.
52. Swanwick, *Thoughts on Education*, 26.
53. "Hints on Reading," *Lady's Magazine* (March 1793), 171.
54. *State Gazette of South Carolina* (May 17, 1773).
55. "On the Supposed Superiority of the Masculine Understanding," 10.
56. "On the Independence and Dignity of the Female Sex," *New York Magazine* (April 1796), 202–203.
57. "Outline of a Plan of Instruction for the Young of both Sexes" (August 11, 1798), 37–41. See also John Hobson, *Prospectus of a Plan of Instruction for the Young of Both Sexes, Including A Course of Liberal Education for Each* (Philadelphia: D. Hogan, printer, 1799).
58. Erasmus Darwin, *A Plan for the Conduct of Female Education in Boarding Schools* (London: J. Drewry, 1797); cited in Catherine Clinton, "Equally Their Due: The Education of the Planter Daughter in the Early Republic," *Journal of the Early Republic* (April 1982), 49.
59. Norton, *Liberty's Daughters*, 235.
60. "An Address to the Ladies," 244.
61. Alexander, *The History of Women*, 66.
62. "Characteristic Differences of the Male and Female of the Human Species," 337.
63. "On Female Authorship," *Lady's Magazine* (January 1793), 72; emphasis in original.
64. Randall, *A Letter*, 12.
65. Mary Hays, "On the Independence and Dignity of the Female Sex," *New York Magazine* (April 1796), 203.
66. For a discussion of the issue of female education and intellect in the magazine *Port Folio*, see William C. Dowling, *Literary Federalism in the Age of Jefferson* (Columbia: University of South Carolina Press, 1999), 80–81. Dowling discusses what he calls "Federal feminism" as expressed in that journal.

67. Virtually every historical account of this period explains that an ideology of "republican motherhood" fueled the advance of women's education. The claim of "republican motherhood" as the principal justification for female education stems from the passage by Rush quoted below. Historians—most notably, Linda K. Kerber—used this quotation as the springboard for conceiving a role for women in the new republic that attached political import to motherhood. See Kerber, *Women of the Republic*.
68. Ibid., 68.
69. Benjamin Rush, "Thoughts Upon Female Education, Accommodated to the Present State of Society, Manners, and Government in the United States of America" [1787], in *Essays on Education in the Early Republic*, Rudolph, ed. (Cambridge: Belknap Press of Harvard University Press, 1965), 28.
70. For further discussion on the limitations of the paradigm of "republican motherhood," see Margaret A. Nash, "Rethinking Republican Motherhood: Benjamin Rush and the Young Ladies' Academy of Philadelphia," *Journal of the Early Republic* 17 (Summer 1997), 171–192.
71. Ruth H. Bloch, "The Gendered Meanings of Virtue in Revolutionary America," 46. Jan Lewis rightly notes that Kerber herself, who coined the phrase under discussion, described women's other roles at least as much as she did motherhood. Lewis remarks that the term "republican motherhood" has "taken on a life of its own and is often assumed to say more about motherhood than Kerber herself ever claimed." Lewis suggests that the term "republican *woman*hood" is far more reflective of the rhetoric of the period. See Jan Lewis, "Republican Wife," 690, fn. 2.
72. Linda K. Kerber, "Daughters of Columbia: Educating Women for the Republic, 1787–1805," in *The Hofstadter Aegis: A Memorial*, Stanley Elkins and Eric McKitrick, eds. (New York: Alfred A. Knopf, 1974), 41–43.
73. For more on this point, see William Casement, "Learning and Pleasure: Early American Perspectives," *Educational Theory* 40 (Summer 1990), 343–349.
74. Samuel Magaw, *An Address Delivered in the Young Ladies Academy, at Philadelphia, on February 8th, 1787. At the Close of a public Examination* (Philadelphia: Thomas Dobson, 1787), 5.
75. "On Female Education," 570.
76. "Reflections on Women," 90.
77. "On the Study of the Arts and Sciences," *New York Magazine* (June 1795), 363–364.
78. Hannah Adams, *A Memoir of Miss Hannah Adams, Written by Herself with Additional Notices by a Friend* (Boston: Gray and Bowen, 1832), 7.
79. Letter from Jonathan Steele to Ann Steele, December 27, 1800, Steele Family Papers, Southern Historical Collection, UNC.
80. Letter from Joyce Myers to Rachel and Ellen Mordecai, December 10, 1796, Jacob Mordecai Papers, Duke University Special Collections.
81. Mary Kelley, " 'A More Glorious Revolution': Women's Antebellum Reading Circles and the Pursuit of Public Influence," *New England Quarterly* 76 (June 2003), 163–196.

82. Address by Benjamin Say, 1789, *The Rise and Progress of the Young Ladies' Academy of Philadelphia*, 30–37; "Essay on Education," *New York Magazine* (January 1790), 40; Address by Dr. Andrews [1788], in *The Rise and Progress of the Young Ladies' Academy of Philadelphia*, 18–21; "Tis Education Forms the Female Mind," 93.

83. Gregory, "Miscellaneous Observations," 162.

84. Valedictory oration, Ann Negus, 1794, in Neal, *Education and Genius of the Female Sex*, 35, 36.

85. "On Female Education," 570.

86. Ann Negus, "Oration," in Neal, *Education and Genius of the Female Sex*, 35.

87. "A Plan for a Matrimonial Lottery," *Lady's Magazine* (June 1792), 173–174.

88. Hays, "On the Independence and Dignity of the Female Sex," 203.

89. On the naming of Southern women as executors of estates, see Thomas Woody, *A History of Women's Education in the United States*, I (New York: The Science Press, 1929), 249, 257.

90. "Female Influence," 300. There are many examples of such arguments for female education. See, for instance, Hobson, *Prospectus of a Plan of Instruction for the young of Both Sexes*; Benjamin Say, "Address," in *The Rise and Progress of the Young Ladies' Academy of Philadelphia*, 31–33; "Outline of a Plan of Instruction for the Young of Both Sexes," *Weekly Magazine* (August 11, 1798), 37–41.

91. Murray, *The Gleaner*, 726–731; emphasis in original.

92. Murray, *The Gleaner*, 726–731.

93. "Improvements Suggested in Female Education," *Monthly Magazine* (March 1797); reprinted in *New York Magazine* (August 1797), 405–408.

94. "The Propriety of Meliorating the Condition of Women in Civilized Societies, Considered," *American Museum* 6 (1789), 248–249.

95. "A Hint," *American Museum* 7 (1790), 208.

96. "Thoughts on Old Maids," *Lady's Magazine* (July 1792), 60.

97. Lord Kaims, *Loose Hints upon Education*, excerpted in "Instructions Preparatory to the Married State," *New York Magazine* (July 1797), 374.

98.      No ties shall perplex, no fetters shall bind,
     That innocent freedom that dwells in my mind.
     At liberty's spring such droughts I've imbibed,
     That I hate all the doctrines by wedlock prescrib'd.

     [R]ound freedom's fair standard I've rallied and paid,
     A Vow of Allegiance to die an old maid.

"Lines, Written by a Lady, Who was Questioned Respecting her Inclination to Marry," *Massachusetts Magazine* 6 (September 1794), 566; quoted in Lee Virginia Chambers-Schiller, *Liberty, A Better Husband: Single Women in America: The Generations of 1780–1840* (New Haven, CT and London: Yale University Press, 1984), 13.

99. Chambers-Schiller, *Liberty, A Better Husband*.

## 3 "Cultivating the Powers of *Human Beings*"

1. Elizabeth Hamilton, *Letters on Education* (Dublin: H. Colbert, 1801), 15, 28–29; emphasis in original.

2. Kim Tolley, "Mapping the Landscape of Higher Schooling, 1727–1850," in *Chartered Schools: Two Hundred Years of Independent Academies in the United States, 1727–1925*, Nancy Beadie and Kim Tolley, eds. (New York and London: RoutledgeFalmer, 2002), 24–26. Also see Thomas Woody, *A History of Women's Education in the United States*, I (New York: The Science Press, 1929), 152–154, 271, 281.

3. Woody, *A History of Women's Education*, I, ch. VIII; Robert Church, *Education in the United States: An Interpretive History* (New York: The Free Press, 1976), ch. 2; Barbara Miller Solomon, *In the Company of Educated Women: A History of Women and Higher Education in America* (New Haven, CT and London: Yale University Press, 1985), ch. 2.

4. John Locke, *Some Thoughts Concerning Education*, quoted in Lorraine Smith Pangle and Thomas L. Pangle, *The Learning of Liberty* (Lawrence: University Press of Kansas, 1993), 54–56.

5. Quoted in Ann Gordon, "The Young Ladies Academy of Philadelphia," in *Women of America: A History*, Carol Ruth Berkin and Mary Beth Norton, eds. (Boston: Houghton Mifflin, 1979), 77.

6. Samuel Knox, "An Essay on the Best System of Liberal Education, Adapted to the Genius of the Government of the United States," in *Essays on Education in the Early Republic*, Frederick Rudolph, ed. (Cambridge: Belknap Press of Harvard University Press, 1965), 305; Pangle and Pangle, *The Learning of Liberty*, 91.

7. Knox, "An Essay on the Best System of Liberal Education," 306.

8. Church, *Education in the United States*, 27–29.

9. See Woody, *A History of Women's Education*, I, ch. VIII; Church, *Education in the United States*, ch. 2; Solomon, *In the Company of Educated Women*, ch. 2; Theodore R. Sizer, "The Academies: An Interpretation," *The Age of the Academies* (New York: Teachers College, 1964), 1–11; E. Merton Coulter, "The Ante-Bellum Academy Movement in Georgia," *Georgia Historical Quarterly* V (December 1921), 11–42; Charles Lee Smith, *The History of Education in North Carolina* (Washington: Government Printing Office, 1888), 45; George Frederick Miller, *The Academy System of the State of New York* (New York: Arno Press & The New York Times, 1969), 25.

10. Benjamin Franklin, "Proposals Relating to the Education of Youth in Pensilvania," [1749]; 68–76; Sizer, "The Academies: An Interpretation," 5–9; both in Sizer, *The Age of the Academies*.

11. Caroline Winterer, *The Culture of Classicism: Ancient Greece and Rome in American Intellectual Life, 1780–1910* (Baltimore and London: The Johns Hopkins University Press, 2002), ch. 1.

12. Noah Webster, "On the Education of Youth in America" [1790], in *Essays on Education in the Early Republic*, Rudolph, ed. 45. Also see Carl J. Richard, *The*

*Founders and the Classics: Greece, Rome, and the American Enlightenment* (Cambridge: Harvard University Press, 1994), 196–231, for a more detailed discussion on the opposition to the classics.

13. Webster, "On the Education of Youth in America," 54.

14. Church, *Education in the United States*, 30–32; Sizer, *The Age of the Academies*, 5–6.

15. David W. Robson, *Educating Republicans: The College in the Era of the American Revolution, 1750–1800* (Westport, CT and London: Greenwood Press, 1985), 204, 208.

16. Church, *Education in the United States*, 23–24, 30.

17. Ibid., 25.

18. Henry Barnard suggested that nine times as many students went to academies as to colleges in 1850; Henry Barnard, "Educational Statistics of the United States in 1850," *American Journal of Education*, I (1855), 368.

19. Solomon, *In the Company of Educated Women*, ch. 4.

20. Jean Pond, *Bradford: A New England Academy* (Bradford, MA: Alumnae Association, 1930), 37–38.

21. Woody, *A History of Women's Education*, I, 299; Cynthia A. Kierner, *Beyond the Household: Women's Place in the Early South, 1700–1835* (Ithaca, NY and London: Cornell University Press, 1998), 158.

22. Benjamin Rush, "A Plan for the Establishment of Public Schools and the Diffusion of Knowledge in Pennsylvania; To Which Are Added, Thoughts upon the Mode of Education, Proper in a Republic," in *Essays on Education in the Early Republic*, Rudolph, ed. 16–17.

23. Church, *Education in the United States*, 37–38.

24. Sizer, "The Academies," 12–13.

25. For additional discussion on this term, see James D. Watkinson, "Useful Knowledge? Concepts, Values, and Access in American Education, 1776–1840," *History of Education Quarterly* 30 (Fall 1990), 351–370.

26. *Massachusetts Magazine* (January 1791), frontispiece.

27. Webster, "On the Education of Youth in America," 70.

28. *The Rise and Progress of the Young Ladies' Academy of Philadelphia: Containing an Account of a Number of Public Examinations & Commencements; The Charter and Bye-Laws; Likewise, A Number of Orations Delivered by the Young Ladies, and Several by the Trustees of said Institution* (Philadelphia: Stewart and Cochran, 1794), 25, 30.

29. Quoted in Winterer, *The Culture of Classicism*, 21.

30. Hannah More, "Thoughts on the Cultivation of the Heart and Temper in the Education of Daughters" [1794], in *The Lady's Companion* (Worcester, MA: The Spy Office, 1824), 92–98.

31. See, for instance, "On Family Ambition," *Lady's Magazine* (August 1792), 121–123. Also see Jan Lewis, "The Republican Wife: Virtue and Seduction in the Early Republic," *William and Mary Quarterly*, 3rd ser., XLIV (October 1987), 689–721; Ruth H. Bloch, "American Feminine Ideals in Transition: The Rise of the Moral Mother, 1785–1815," *Feminist Studies* 4 (June 1978), 117.

32. Maria Edgeworth and Richard Lovell Edgeworth, *Practical Education* [1798] (New York: George F. Hopkins, 1801), 117.

33. Franklin, "Proposals Relating to the Education of Youth," 70. Ever the pragmatist, Franklin also said that "art is long and life is short," and therefore students should prioritize practical subjects over the arts. However, this is not a particularly gendered argument, since Franklin saw benefits for both men and women in learning the arts.

34. Simeon Doggett, "A Discourse on Education, Delivered at the Dedication and Opening of Bristol Academy, the 18th Day of July, A.D. 1796," in *Essays on Education in the Early Republic*, Rudolph, ed., 152.

35. Samuel Smith, "Remarks on Education: Illustrating the Close Connection Between Virtue and Wisdom. To Which Is Annexed a System of Liberal Education," in *Essays on Education in the Early Republic*, Rudolph, ed., 217–218.

36. Knox, "An Essay on the Best System of Liberal Education," 356.

37. John Adams to Abigail Adams, August 28, 1774; quoted in Edith B. Gelles, *Portia: The World of Abigail Adams* (Bloomington: Indiana University Press, 1992), 143; quoted in Gelles, *Portia*, 136.

38. Information on curricula comes from the following sources: *Laws of the Raleigh Academy: With the Plan of Education Annexed* (Raleigh: Gales & Seaton, 1811); Jacob Mordecai Papers, Duke University Special Collections; Jean Pond, *Bradford: A New England Academy*; John Swanwick, *Thoughts on Education, Addressed to the Visitors of the Young Ladies' Academy in Philadelphia, October 31, 1787* (Philadelphia: Thomas Dobson, 1787); *Terms and Conditions of the Boarding School for Female Education in Salem, N.C.*, Broadside, December 1807, North Carolina Collection, University of North Carolina-Chapel Hill; *The Rise and Progress of the Young Ladies' Academy of Philadelphia*; *Young Ladies' Boarding-School, Warrenton*, Broadside, May 19, 1809, Duke University Special Collections; Emily Vanderpoel, *Chronicles of a Pioneer School, from 1792 to 1833* (Cambridge, MA: Cambridge University Press, 1903), and Vanderpoel, *More Chronicles of a Pioneer School, from 1792 to 1833* (New York: The Cadmus Book Shop, 1927); Emory Washburn, *Brief Sketch of a History of Leicester Academy* (Boston: Phillips, Sampson, 1855); Woody, *A History of Women's Education*, I, chs. 4–6; *North Carolina Minerva and Fayetteville Advertiser* (June 3, 1797), 3; (June 30, 1798), 3; (July 7, 1798), 3; *Raleigh Register* (September 10, 1813); *State Gazette of North Carolina* (January 22, 1789), 3; *State Gazette of South Carolina* (November 11, 1790), 4; (November 18, 1790), 3; (January 13, 1791), 2; (February 7, 1791), 1; (April 21, 1791), 1.

39. Benjamin Rush, "Thoughts Upon Female Education, Accommodated to the Present State of Society, Manners, and Government in the United States of America," in *Essays on Education in the Early Republic*, Rudolph, ed., 29.

40. Rollo Laverne Lyman, *English Grammar in American Schools Before 1850* (Washington: Government Printing Office, 1922), 39–43; Edwin C. Broome, *A Historical and Critical Discussion of College Admission Requirements* (New York: Macmillan & Co., 1903), 43; Robson, *Educating Republicans*, 150.

41. John Teaford, "The Transformation of Massachusetts Education, 1670–1780," in *The Social History of American Education*, B. Edward McClellan and William J. Reese, eds. (Urbana and Chicago: University of Illinois Press, 1988), 23–38; Kenneth Lockridge, *Literacy in Colonial New England: An Enquiry into the Social Context of Literacy in the Early Modern West* (New York: W. W. Norton & Co., 1974), 127.
42. Rush, "Thoughts Upon Female Education," 29.
43. *The Rise and Progress of the Young Ladies' Academy of Philadelphia*, 16.
44. Swanwick, "Thoughts on Education," 9.
45. "The System of Public Education, Adopted by the Town of Boston, October 15, 1789," *New York Magazine* (January 1790), 52; Woody, *A History of Women's Education*, I, 299. The names "reading school" and "writing school" are misleading: Boston's reading school taught spelling, accent, reading, grammar, and composition, while the writing school taught writing and arithmetic (through fractions).
46. Kierner, *Beyond the Household*, 153.
47. Samuel Magaw, *An Address Delivered in the Young Ladies Academy, at Philadelphia, on February 8th, 1787. At the Close of a Public Examination* (Philadelphia: Thomas Dobson, 1787), 10. Samuel Knox's plan for education also included instruction for both sexes in, not only "mere pronunciation, but also the management of the voice with gracefulness and propriety." See Knox, "An Essay on the Best System of Liberal Education," 329.
48. Woody, *A History of Women's Education*, I, 340.
49. *The Rise and Progress of the Young Ladies' Academy of Philadelphia*, 38.
50. Ibid., 73–75.
51. Pangle and Pangle, *The Learning of Liberty*, 82.
52. Woody, *A History of Women's Education*, I, 412.
53. Ibid.
54. Ibid., 29.
55. *The Rise and Progress of the Young Ladies' Academy of Philadelphia*, 32.
56. Woody, *A History of Women's Education*, I, 156. The Rule of Three was as follows. (1) When the quantity of one commodity is given with its value, or the value of its integer, as also the value of the integer of some other commodity to be exchanged for it, to find the quantity of this commodity: Find the value of the commodity of which the quantity is given, then find how much of the other commodity, at the rate proposed, may be had for that sum. (2) If the quantities of both commodities be given, and it should be required to find how much of some other commodity, or how much money should be given for the inequality of their values: Find the separate values of the two given commodities, subtract the less from the greater, and the remainder will be the balance, or value of the other commodity. (3) If one commodity is rated above the ready money price, to find the bartering price of the other: Say, as the ready money price of the one, is to the bartering price, so is that of the other to its bartering price. See Church, *Education in the United States*, 14–15.
57. "Improvements Suggested in Female Education," *New York Magazine* (August 1797), 407.

58. Kim Tolley, *The Science Education of American Girls: A Historical Perspective* (New York and London: RoutledgeFalmer, 2003), 22.

59. Jedediah Morse, *Geography Made Easy* [1784] (Boston: J. T. Buckingham, 1806); see John A. Nietz, *Old Textbooks* (Pittsburgh: University of Pittsburgh Press, 1961), 217–218.

60. Tolley, *The Science Education of American Girls*, 30–31; William J. Reese, *The Origins of the American High School* (New Haven, CT and London: Yale University Press, 1995), 118.

61. Clifton Johnson, *Old-Time Schools and School-Books* (Gloucester, MA: Peter Smith, 1963), 320.

62. Anya Jabour, *Marriage in the Early Republic: Elizabeth and William Wirt and the Companionate Ideal* (Baltimore and London: The Johns Hopkins University Press, 1998), 11–12.

63. *Richmond Enquirer*, January 9, 1813; May 14, 1813; August 27, 1819; quoted in Kierner, *Beyond the Household*, 151.

64. Joan W. Goodwin, *The Remarkable Mrs. Ripley: The Life of Sarah Alden Bradford Ripley* (Boston: Northeastern University Press, 1998), 14–15.

65. Letter from Rachel Mordecai to Caroline Mordecai, February 19, 1812, Mordecai Family Papers, Duke Special Collections, Duke University.

66. Tolley, "Mapping the Landscape," 33.

67. "On Female Authorship," *Lady's Magazine* 1 (1793), 69.

68. Winterer, *The Culture of Classicism*, 24–29.

69. Mary Beth Norton, *Liberty's Daughters: The Revolutionary Experience of American Women, 1750–1800* (New York: HarperCollins, 1980), 23.

70. Jeanne Boydston, *Home & Work: Housework, Wages, and the Ideology of Labor in the Early Republic* (New York and Oxford: Oxford University Press, 1990), 41.

71. Norton, *Liberty's Daughters*, 24; Boydston, *Home & Work*, 37.

72. M.F.B., "Answer to a Father's Inquiries relative to the Education of Daughters," *The New England Quarterly Magazine* (December 1802), 156. Lynn Templeton Brickley also notes that the practical application of needlework cut across all socioeconomic and class boundaries: poor women earned their livings as seamstresses, middle-class women taught needlework, and wealthy women supervised the creation and maintenance of their household's clothing, linens, and furnishings. See Brickley, "Sarah Pierce's Litchfield Female Academy, 1792–1833," Ph.D. Dissertation, Harvard University, 1985, 160–161.

73. John Cossens Ogden, *The Female Guide* (Concord, NH: George Hough, 1793), 27; Swanwick, *Thoughts on Education*, 22.

74. Brickley, "Sarah Pierce's Litchfield Female Academy," 163.

75. Rush, "Plan for the Establishment of Public Schools," 16.

76. Rush, "Thoughts Upon Female Education," 34.

77. Edgeworth and Edgeworth, *Practical Education*, 118.

78. Ibid.

79. Thomas Woody's findings corroborate this pattern. See Woody, *A History of Women's Education*, I, 418.

80. Hamilton, *Letters on Education*, 15, 28–29.

81. Church, *Education in the United States*, 34; Carl F. Kaestle, *Pillars of the Republic: Common Schools and American Society, 1780–1860* (New York: Hill and Wang, 1983), 18, 45–46, 97.
82. "Outline of a Plan of Instruction for the Young of Both Sexes, particularly Females, submitted to the Reflection of the Intelligent and the Candid," *Weekly Magazine of Original Essays, Fugitive Pieces, and Interesting Intelligence* (August 11, 1798), 39; emphasis in original.
83. "Hints on Reading," *Lady's Magazine* (March 1793), 172.
84. Edgeworth and Edgeworth, *Practical Education*, 143.
85. Solomon, *In the Company of Educated Women*, 17–18.
86. See chapter 5 in this book for a discussion of this.
87. *The Rise and Progress of the Young Ladies' Academy of Philadelphia*, 27, 34, 51–52.
88. Swanwick, "Thoughts on Education," 7–8.
89. See, for instance, Edgeworth and Edgeworth, *Practical Education*, 141.

## 4 FEMALE EDUCATION AND THE EMERGENCE OF THE "MIDDLING CLASSES"

1. Maria Budden, *Thoughts on Domestic Education; The Result of Experience. By a Mother* (London: Charles Knights, 1826), 71, 74–75. The *American Journal of Education* reprinted portions of this book in 1828.
2. One study found that over 500 colleges, primarily male, were founded in the antebellum era (approximately 80% of which no longer existed by the 1920s), while another scholar estimates well over 700. See Donald G. Tewksbury, *The Founding of American Colleges and Universities Before the Civil War* (New York: Arno Press & The New York Times, 1969), 27–29; Frederick Rudolph, *The American College and University: A History* (New York: Alfred A. Knopf, 1968), 47. Hundreds of institutions for the higher schooling of women were also founded during these years; see Thomas Woody, *A History of Women's Education in the United States*, I (New York: The Science Press, 1929), ch. VIII.
3. See, for instance, Stuart M. Blumin, *The Emergence of the Middle Class: Social Experience in the American City, 1760–1900* (Cambridge: Cambridge University Press, 1989); Michael B. Katz, *The Social Organization of Early Industrial Capitalism* (Cambridge: Cambridge University Press, 1982); Mary P. Ryan, *The Cradle of the Middle-Class: The Family in Oneida County, New York, 1780–1865* (Cambridge: Cambridge University Press, 1981).
4. Blumin, *The Emergence of the Middle Class*, 245. Also see Nancy Beadie's discussion regarding the difficulty of placing academy students into the category of the middle class. Nancy Beadie, "'To Improve Every Leisure Moment': The Significance of Academy Attendance in the Mid-19th Century," Paper Presented at the History of Education Society annual meeting, October, 1999.
5. Some classic works on this topic are Paul E. Johnson, *A Shopkeeper's Millennium: Society and Revivals in Rochester, New York, 1815–1837* (New York: Hill and Wang,

1978); Nancy A. Hewitt, *Women's Activism and Social Change: Rochester, New York, 1822–1872* (Ithaca, NY and London: Cornell University Press, 1984). See also Anne M. Boylan, *The Origins of Women's Activism: New York and Boston, 1797–1840* (Chapel Hill and London: The University of North Carolina Press, 2002).

6. See William J. Reese, *The Origins of the American High School* (New Haven, CT and London: Yale University Press, 1995), chs. 2 and 3; Carl F. Kaestle, *Pillars of the Republic: Common Schools and American Society, 1780–1860* (New York: Hill and Wang, 1983), chs. 4–6.

7. Barbara Welter, "The Cult of True Womanhood," *American Quarterly* 18 (Summer 1966), 151–174; Nancy F. Cott, *The Bonds of Womanhood: "Woman's Sphere" in New England, 1780–1835* (New Haven, CT: Yale University Press, 1977).

8. Many historians have made this point. See Boylan, *The Origins of Women's Activism*; Nancy A. Hewitt, *Women's Activism and Social Change*; Mary P. Ryan, *The Cradle of the Middle-Class*; Jeanne Boydston, Mary Kelley, and Anne Margolis, *The Limits of Sisterhood: The Beecher Sisters on Women's Rights and Woman's Sphere* (Chapel Hill and London: University of North Carolina Press, 1988), especially the Introduction.

9. Catherine A. Brekus, *Strangers & Pilgrims: Female Preaching in America, 1740–1845* (Chapel Hill and London: University of North Carolina Press, 1998), 123.

10. Ibid., 143, 286–287.

11. Rudolph, *The American College and University*, chs. 3 and 4; Tewksbury, *The Founding of American Colleges and Universities Before the Civil War*, ch. 2.

12. Discussions of evangelicalism abound. For examples, see Johnson, *A Shopkeeper's Millennium*; William G. McLoughlin, *Revivals, Awakenings, and Reform: An Essay on Religion and Social Change in America, 1607–1977* (Chicago: University of Chicago Press, 1978); Caroll Smith-Rosenberg, *Religion and the Rise of the American City: The New York City Mission Movement, 1812–1820* (Ithaca NY: Cornell University Press, 1971).

13. Barbara Leslie Epstein, *The Politics of Domesticity: Women, Evangelism, and Temperance in Nineteenth Century America* (Middletown CT: Wesleyan University Press, 1981), 45.

14. See Cott, *The Bonds of Womanhood*, especially ch. 4; Lori D. Ginzberg, *Women and the Work of Benevolence: Morality, Politics, and Class in the 19th-Century United States* (New Haven, CT and London: Yale University Press, 1990), especially ch. 1.

15. "Introductory Lecture at the Seminary," Byfield, May 1821, Joseph Emerson Papers, MHC; "Lectures on Intellectual Philosophy, delivered at Ipswich," Ipswich Female Seminary, 1834, MHC.

16. James M. Garnett, *Lectures on Female Education, Comprising the First and Second Series of a Course Delivered to Mrs. Garnett's Pupils, At Elm-Wood, Essex County, Virginia. By James M. Garnett. To Which is Annexed, The Gossip's Manual* (Richmond, VA: Thomas W. White, 1825), 296.

17. John Ruggles Cotting, *An Address Delivered at the Female Classical Seminary, Brookfield, Jan. 22, 1827, at the Close of the Introductory Lecture at the Winter*

*Course on Natural Science* (Brookfield, MA: E. and G. Merriam, 1827), 6. For other examples of this view, see Joseph Story, *A Discourse pronounced before the Phi Beta Kappa Society, at the Anniversary Celebration, on the Thirty-first Day of August, 1826* (Boston: Hilliard, Gray, Little, and Wilkins, 1826), 16; "Introductory Lecture at the Seminary," Byfield, May 1821, Joseph Emerson Papers, MHC; "Some Remarks on the General Object of Education," attributed to Mary Lyon, Buckland Female School, Nov. 11, 1829, MHC.

18. *Catalogue of the Officers and Members of Granville Female Academy, for the year ending February 22, 1838.* Granville, Ohio (Columbus, OH: Cutler and Pilsbury, 1838), 13.

19. James M. Garnett, "An Address on the Subject of Literary Associations to Promote Education; Delivered before the Institute of Education of Hampden Sidney College, Va., at Their Last Commencement," in *American Annals of Education and Instruction* V (July 1835), 317.

20. Amherst Academy circular, June 1832, in Amherst Academy Records (Box 1, Folder 4), Amherst College Archives.

21. "Education of Females," *American Journal of Education* II (August 1827), 485.

22. *General View of the Plan of Education Pursued at the Adams Female Academy* (Exeter, NH: Nathaniel S. Adams, printer, 1831), 3.

23. *Catalogue of the Alabama Female Institute, Tuscaloosa, Ala. for the Year Ending July, 1838* (Tuscaloosa, AL: The Intelligencer Office, 1838), 8–9 (emphasis in original). There are numerous examples of the emphasis on moral education in female seminaries and academies; a few of them include *Ballston Spa Female Seminary* (Albany, NY: Packard & Van Benthuysen, 1824), 4; Cotting, *An Address Delivered at the Female Classical Seminary, Brookfield*, 3; *Annual Circular, Etc., of the Geneva Female Seminary, under the Care of Mrs. Ricord* (Geneva: Merrill, printer, 1840), 7–8; Rev. John W. Scott, *An Address on Female Education, Delivered at the Close of the Summer Session for 1840, of the Steubenville Female Seminary, in Presence of its Pupils and Patrons* (Steubenville, OH: n.p., 1840), 4.

24. For men, see Rudolph, *The American College and University*, 74. For women, see catalogs and circulars listed in the appendix in this book.

25. Rudolph, *The American College and University*, 140–141.

26. See the appendix for a list of school catalogs and circulars used in this study.

27. Charles Burroughs, *An Address on Female Education, Delivered in Portsmouth, New-Hampshire, October 26, 1827* (Portsmouth: Childs and March, 1827), 26.

28. See Cott, *The Bonds of Womanhood*, 84–85.

29. "On the Policy of Elevating the Standard of Female Education, Addressed to the American Lyceum, May, 1834," *American Annals of Education and Instruction* IV (August 1834), 362.

30. Daniel Mayes, *An Address Delivered on the First Anniversary of Van Doren's Collegiate Institute for Young Ladies, in the City of Lexington, Ky. On the Last Thursday of July, 1832* (Lexington, KY: Finnell & Herndon, 1832), 12. References to the importance of education in the preparation of good mothers are numerous. For more examples, see Joseph Emerson, *Female Education. A Discourse, Delivered at the Dedication of the Seminary Hall in Saugus, Jan. 14,*

*1822* (Boston: Samuel T. Armstrong, and Crocker & Brewster, 1823), 8; *Ballston Spa Female Seminary*, 4; Garnett, *Lectures on Female Education*, 8, 12, 22; John T. Irving, *Address Delivered on the Opening of the New-York High-School for Females, January 31, 1826* (New York: William A. Mercein, 1826), 12–13; Burroughs, *An Address on Female Education*, 6.

31. "Prospectus," *American Journal of Education* I (January 1826), 3.

32. "Destination of Woman," originally printed in *Charleston Observer*, reprinted in *American Annals of Education and Instruction* VI (November 1836), 503; "Female Education," *American Journal of Education* III (September 1828), 519. For other examples, see *American Journal of Education* III (January 1828), 55; (March 1828), 183; *American Annals of Education*: "Progress of Female Education" (September 1830), 97; "Fundamental Principles of Female Education" (February 1834), 86; "Female Education" (October 1837), 449.

33. Garnett, *Lectures on Female Education*, 22. See also, Elias Marks, *Hints on Female Education, with an Outline of an Institution for the Education of Females, Termed the So. Cal. Female Institute; under the Direction of Dr. Elias Marks* (Columbia, SC: David W. Sims, printer, 1828), 9.

34. Emerson, *Female Education*, 4–5, 10, 12–13.

35. Abigail Mott, *Observations on the Importance of Female Education, and Maternal Instruction, with their Beneficial Influence on Society. By a Mother* (New York: Mahlon Day, printer, 1825), 13; Samuel M. Burnside, "Speech on the Opening of Two New Female Schools" (Worcester, MA: n.p., 1833?). See also Daniel Chandler, *An Address on Female Education, Delivered before the Demosthenian & Phi Kappa Societies, on the Day after Commencement, in the University of Georgia* (Washington, GA: William A. Mercer, 1835), 14; *Prospectus of the Raleigh Academy, and Mrs. Hutchison's View of Female Education* (Raleigh, NC: C. White, printer, 1835), 6.

36. Marks, *Hints on Female Education*, 9.

37. *Catalogue of the Young Ladies' Seminary, in Keene, N.H. for the Year Ending October, 1832* (Keene, NH: J. & J. W. Prentiss, 1832), 11.

38. Martha Hazeltine, "Letter from the Corresponding Secretary," *Third Annual Report of the Young Ladies' Association of the New-Hampton Female Seminary, for the Promotion of Literature and Missions; with the Constitution, etc. 1835–36* (Boston: John Putnam, 1837), 42.

39. Carroll Smith-Rosenberg, "Beauty, the Beast and the Militant Woman: A Case Study in Sex Roles and Social Stress in Jacksonian America," *American Quarterly* 23 (October 1971), 562–584.

40. Hewitt, *Women's Activism and Social Change*, 41.

41. Cynthia A. Kierner, *Beyond the Household: Women's Places in the Early South, 1700–1835* (Ithaca, NY and London: Cornell University Press, 1998), 184–192.

42. "Female Education in the Last Century," *American Annals of Education and Instruction* I (November 1831), 526.

43. Maria Cowles to Henry Cowles, March 29, 1831, Ipswich Students, Alumnae, and Teachers, 1830–1863, Correspondence, Series E, Folder 2, Mount Holyoke College Archives.

44. Catharine Beecher, *An Essay on the Education of Female Teachers* (New York: Van Nostrand & Dwight, 1835). For a discussion of this essay, see Kathryn Kish Sklar, *Catharine Beecher: A Study in American Domesticity* (New York and London: W. W. Norton & Company, 1976), 113f.; Boydston, Kelley, and Margolis, *The Limits of Sisterhood*, 115f.

45. Zilpah Grant, "Benefits of Female Education," 1836, MHCA.

46. *Third Annual Report of the Young Ladies' Association of the New-Hampton Female Seminary*, 20, 23.

47. *Oberlin Collegiate Institute: 1st Circular, March 8, 1834* (n.p., 1834).

48. Many historians cite these factors. See Joel Perlman and Robert A. Margo, *Women's Work? American Schoolteachers, 1650–1920* (Chicago: University of Chicago Press, 2001), 86–109; John L. Rury, "Who Became Teachers and Why: The Social Characteristics of Teachers in American History," in *American Teachers: Histories of a Profession at Work*, Donald Warren, ed. (New York: Macmillan Publishing Company, 1989), 9–48; Geraldine Jonich Clifford, "Man/Woman/Teacher: Gender, Family and Career in American Educational History," in *American Teachers*, Donald Warren, ed. 293–343; Michael W. Apple, "Teaching and 'Women's Work': A Comparative Historical and Ideological Analysis," *Teachers College Record* 86 (Spring 1985), 457–473; Sklar, *Catharine Beecher*, 97–98; Barbara Miller Solomon, *In the Company of Educated Women: A History of Women and Higher Education in America* (New Haven, CT and London: Yale University Press, 1985), ch. 2; Kaestle, *Pillars of the Republic*, ch. 6; David B. Tyack and Myra H. Strober, "Jobs and Gender: A History of the Structuring of Educational Employment by Sex," in *Educational Policy and Management: Sex Differentials*, Patricia Schmuck, ed. (New York: Academic Press, 1981), 136.

49. "Prospectus," 1.

50. "Education in the State of New-York: Extract from Gov. Clinton's Message, Jan. 3, 1826," *American Journal of Education* I (January 1826), 58.

51. Reese, *The Origins of the American High School*, 21–24. For an overview of the growth of the common school movement, see Kaestle, *Pillars of the Republic*.

52. "Miss Beecher's Essay on the Education of Female Teachers," *American Annals of Education and Instruction* V (June 1835), 277.

53. "Domestic Seminary for Young Ladies," *American Annals of Education and Instruction* IV (November 1834), 499.

54. Quoted in Clarence P. McClelland, *The Education of Females in Early Illinois* (Jacksonville, IL: MacMurray College for Women, 1944), 13.

55. Richard M. Bernard and Maris A. Vinovskis, "The Female School Teacher in Ante-Bellum Massachusetts," *Journal of Social History* 10 (March 1977), 337.

56. Hannah H. Barker to Mr. Shipherd, July 1, 1835; Sally Eliza Nash, to Mrs. Eliza Stewart, February 1836; in Treasurer's Office Correspondence, 1822–1907, Oberlin College Archives.

57. David F. Allmendinger, Jr., "Mount Holyoke Students Encounter the Need for Life-Planning, 1837–1850," *History of Education Quarterly* 19 (Spring 1979), 28–29.

58. Bernard and Vinovskis, "The Female School Teacher in Ante-Bellum Massachusetts," 332; Jo Anne Preston, "Domestic Ideology, School Reformers, and Female Teachers: Schoolteaching Becomes Women's Work in

Nineteenth-Century New England," *The New England Quarterly* LXVI (December 1993), 531–532. Of course, being willing to work for lower pay is not the same as liking to do so, and some women objected to the pay differential between men and women. For one example of someone who resented the inequity, see Andrea Moore Kerr, *Lucy Stone: Speaking Out for Equality* (New Brunswick, NJ: Rutgers University Press, 1992), ch. 3.

59. Emerson, *Female Education*, 11. A speaker at the American Lyceum echoed this, saying that "females are by nature designated as teachers," and "to teach is their province." See "On the Policy of Elevating the Standard of Female Education," 362.

60. "Self Improvement an Important Part of Female Education," *American Journal of Education* III (March 1828), 161–162.

61. Robert L. Church, *Education in the United States: An Interpretive History* (New York: The Free Press, 1976), 78.

62. See, for instance, Lewis Perry, *Intellectual Life in America* (New York: Franklin Watts, 1984), ch. 4; Richard D. Brown, *Knowledge Is Power: The Diffusion of Information in Early America, 1700–1965* (New York and Oxford: Oxford University Press, 1989), ch. 9. Although Brown has a chapter on women (Chapter 7: Daughters, Wives, Mothers: Domestic Roles and the Mastery of Affective Information, 1765–1865), his focus there is on communication networks and the means through which women acquired information; it is not on women in the self-improvement movement, the way his chapter 9 is for men. For an example of a treatment of the movement that takes both women and men into account, see Joseph F. Kett, *The Pursuit of Knowledge Under Difficulties: From Self-Improvement to Adult Education in America, 1750–1990* (Stanford, CA: Stanford University Press, 1994).

63. Burroughs, *An Address on Female Education*, 7.

64. "Education of Females: Motives to Application," *American Journal of Education* II (September 1827), 549–550.

65. Cotting, *An Address Delivered at the Female Classical Seminary, Brookfield*, 3.

66. Almira Hart Lincoln Phelps, *Caroline Westerley; or, The Young Traveller from Ohio. Containing the Letters of a Young Lady of Seventeen, Written to Her Sister* (New York: J. & J. Harper, 1833), 36.

67. "While a child has a right in his very nature to be fed, he has the same right to be educated." Isaac Ferris, *Address Delivered 27th April, 1839, At the Opening of the Rutgers Female Institute, New-York* (New York: William Osborn, printer, 1839), 6, 8, 13.

68. "Education of Females: Intellectual Instruction," *American Journal of Education* II (November 1827), 676–677.

69. Leonard Worcester, *An Address on Female Education. Delivered at Newark (N.J.) March 28, 1832* (Newark? n.p., 1832), 2.

70. *Ballston Spa Female Seminary*, 4–5.

71. Martha Patsy Lenoir Pickens to Julia Pickens Howe, February 13, 1828, Chiliab Howe Papers, Southern Historical Collection.

72. Frederick Peck to Sophia Peck, May 24, 1835 (emphasis in original); Sophia Peck to Eliza Peck, November 25, 1835; Henry Watson, Jr. Papers, Duke Special Collections.

73. The young woman's mother also wrote, "I hope you will study to improve every moment of your time." D. Parish to Lydia and Julia Parish, August 24, 1837, Parish Family Papers, Southern Historical Collection.

74. Letter of Martha Whiting to former student, June 22, 1833, quoted in Catharine Badger, *The Teacher's Last Lesson: A Memoir of Martha Whiting, Late of the Charlestown Female Seminary. Consisting Chiefly of Extracts from Her Journal, Interspersed with Reminiscences and Suggestive Reflections* (Boston: Gould and Lincoln, 1855), 120.

75. Catharine Beecher to Mary Lyon, July 10, 1829, quoted in Marion Lansing, ed., *Mary Lyon Through Her Letters* (Boston: Books, Inc., 1937), 11–12.

76. Elizabeth Alden Green, *Mary Lyon and Mount Holyoke: Opening the Gates* (Hanover, NH: University Press of New England, 1979), 36–37.

77. Garnett, *Lectures on Female Education*, 255.

78. Marks, *Hints on Female Education*, 8.

79. Letter from Charlotte Johnson to Asa Mahan, April 7, 1836, Treasurer's Office Correspondence, Oberlin College Archives.

80. Harry L. Watson, *Liberty and Power: The Politics of Jacksonian America* (New York: Hill and Wang, 1990), 28–34; Charles Sellers, *The Market Revolution: Jacksonian America, 1815–1846* (New York: Oxford University Press, 1991).

81. Mrs. Townshend Stith, *Thoughts on Female Education* (Philadelphia: Clark & Raser, printers, 1831), 29; Almira Hart Lincoln Phelps, *The Female Student; Or, Lectures to Young Ladies on Female Education*, 2nd ed. (New York: Leavitt, Lord & Co., 1836), 29–30; Hannah Farnham Sawyer Lee, *The Contrast: or Modes of Education* (Boston: Whipples and Damrell, 1837).

82. "The Widow's Son," *Lowell Offering* 2 (1841), 246–250.

83. "Disasters Overcome," *Lowell Offering* 2 (1841), 289–297.

84. Irving, *Address Delivered on the Opening of the New-York High-School for Females*, 5–9.

85. "Seminary for Female Teachers," *American Annals of Education and Instruction* I (August 1831), 341.

86. "Miss Beecher's Essay on the Education of Female Teachers," 277.

87. President Beecher of Illinois College, 1833, quoted in McClelland, *The Education of Females in Early Illinois*, 13.

88. Emerson, *Female Education*, 11.

89. Joseph Emerson, *Prospectus of the Female Seminary, at Wethersfield, CT, Comprising a General Prospectus, Course of Instruction, Maxims of Education, and Regulations of the Seminary* (Wethersfield, CT: A. Francis, printer, 1826), 55.

90. Catharine Beecher, *Suggestions Respecting Improvements in Education Presented to the Trustees of the Hartford Female Seminary, and published at their Request* (Hartford: Packard & Butler, 1829), 46. In 1829, Beecher advocated opening the profession of teaching to women; by 1846, she argued that teaching should be the special "profession of women." See Beecher, *The Evils Suffered by American Women and American Children: The Causes and the Remedy* (New York: Harper & Bros., 1846), 11.

91. Quoted in McClelland, *The Education of Females in Early Illinois*, 13.

92. The story of the professionalization of teachers has been told in many places. See Donald Warren, ed., *American Teachers: Histories of a Profession at Work* (New York: Macmillan, 1989); Sari Knopp Biklen, *School Work: Gender and the Cultural Construction of Teaching* (New York: Teachers College Press, 1995); David Tyack and Elisabeth Hansot, *Managers of Virtue: Public School Leadership in America, 1820–1980* (New York: Basic Books, 1982); Jurgen Herbst, *And Sadly Teach: Teacher Education and Professionalization in American Culture* (Madison: University of Wisconsin Press, 1989); Paul H. Mattingly, *The Classless Profession: American Schoolmen in the Nineteenth Century* (New York: New York University Press, 1975).

93. "Prefatory Address," *American Journal of Education* I (January 1827), 6.

94. Kaestle, *Pillars of the Republic*, 20.

95. Ibid.

96. The first normal schools were the three that Massachusetts established in 1839, and the concept quickly spread to other states. There were only 12 state normal schools before the Civil War, and well over one hundred by the end of the nineteenth century. Willard S. Elsbree, *The American Teacher: Evolution of a Profession in a Democracy* (New York: American Book Company, 1939), 146, 152; Woody, *A History of Women's Education*, I, 482–483.

97. Lansing, *Mary Lyon Through Her Letters*, 45. Joseph Emerson's Wethersfield Academy in Connecticut publicly stated its goal of teaching "pupils to teach themselves and each other," thus preparing them both for teaching and for lifelong learning. See Emerson, *Prospectus of the Female Seminary, at Wethersfield*, 17.

98. *An Account of the High School for Girls, Boston: with A Catalogue of the Scholars. February, 1826* (Boston: Thomas B. Wait and Son, printers, 1826), 9–10.

99. *Greenfield Gazette*, May 19, 1828, quoted in *American Journal of Education* III (August 1828), 489. There are numerous examples of institutions that specifically had this goal. For instance, the following notices appeared in the *American Journal of Education*: The Salem, Indiana Female Institute announced that "it is a leading object of Mr M[orrison] to prepare young ladies for teaching" (July 1837, 332); at the Institution for Females at LeRoy, New York, "one principal object is to prepare young ladies to become teachers" (August 1837, 379); the Uxbridge, Massachusetts stated its intention "to prepare young ladies to become teachers" (January 1838, 43). Some of these institutions were coeducational. For instance, the Plymouth, Massachusetts Teachers' Seminary enrolled 110 men and 90 women (January 1838), 43.

100. *Amherst Academy, Mass.* (Amherst: Amherst College, 1827), broadsheet; *Amherst Academy* (Amherst: n.p., 1832), 1.

101. *Catalogue of the Officers and Members of the Ipswich Female Academy* (Ipswich, MA: John Harris, Jr., printer, 1829), 11, 12.

102. *Catalogue of the Officers and Members of the Seminary for Female Teachers, at Ipswich, Mass. for the Year Ending April 1837* (Salem, MA: Palfray and Chapman, printers, 1837). This name change supports Helen Lefkowitz Horowitz's hypothesis that schools used the term "seminary" to connote that it taught women to be teachers in the same way that theological seminaries

trained men to be ministers. See Helen Lefkowitz Horowitz, *Alma Mater: Design and Experience in the Women's Colleges from Their Nineteenth-Century Beginnings to the 1930s* (New York: Alfred A. Knopf, 1984), 11.

103. "Uxbridge Female Seminary" (Worcester, MA: n.p., 1837). New-Hampton Female Institute's stated goal was "to prepare its members either to become teachers or missionaries." See "Female Seminaries," *American Annals of Education and Instruction* VI (May 1836), 235.

104. Letter from Abiah Chapin to Zilpah Grant, November 8, 1836, MHCA; see also *Catalogue of the Officers and Members of the Ipswich Female Academy*, 12.

105. Anne Firor Scott, "The Ever-Widening Circle: The Diffusion of Feminist Values from the Troy Female Seminary, 1822–1872," *History of Education Quarterly* 19 (Spring 1979), 8.

106. "Society for the Education of Females," 1835, MHC Box LD 7093.38 I6— Ipswich Female Seminary, Series D, Folder 1; "Benefits of Female Education," 1836, MHC Box LD 7093.38 I6, College History.

107. *Address Delivered 27th April, 1839, at the Opening of the Rutgers Female Institute, New-York* (New York: William Osborn, printer, 1839), 11. For another example, see *Catalogue of the Officers and Students of Bradford Academy, Bradford, Massachusetts, October, 1839* (Haverhill, MA: E. H. Safford, 1839), 12.

108. *Fourth Annual Report of the Young Ladies' Association of the New-Hampton Female Seminary, for the Promotion of Literature and Missions; with the Constitution, etc. 1837–38* (Boston: Freeman and Bolles, 1838), 22; *Fifth Annual Report of the Young Ladies' Association of the New-Hampton Female Seminary, for the Promotion of Literature and Missions; with the Constitution, etc. 1838–39* (Boston: John Putnam, 1839), 12.

109. Catharine M. Sedgwick, *Means and Ends, or Self-Training* (Boston: March, Capen, Lyon & Webb, 1839), 19, 29–30.

110. "Prospectus of St. Joseph's Academy for Young Ladies" (Broadside: n.p., AAS), 2; *Arcade Ladies' Institute, Providence R.I.* (Providence, RI: H.H. Brown, 1834?), 10; *Annual Catalogue of the Instructers* [sic] *and Pupils in the Newburgh Female Seminary, Situated in Newburgh, Orange Co. N.Y. for the Year preceding January 1st, 1837* (Newburgh, NY: J. D. Spalding, Printer, 1837), 8; *The Western Collegiate Institute, for Young Ladies* (Pittsburgh: n.p., 1837).

111. "Rochester Female Seminary," *Rochester Daily Democrat* (March 31, 1834), 2.

112. Story, *A Discourse pronounced before the Phi Beta Kappa society*, 4–10.

113. Phelps, *Caroline Westerley*, 36.

114. For another example, see Lee, *The Contrast*.

115. "Female Education" (September 1828), 520.

116. Quotes and engravings appear in Milo M. Naeve, *John Lewis Krimmel: An Artist in Federal America* (Newark: University of Delaware Press), 111–114.

117. Kathryn Kish Sklar, "The Founding of Mount Holyoke College," in *Women of America: A History*, Carol R. Burkin and Mary Beth Norton, eds. (Boston: Houghton Mifflin, 1979), 177–201.

118. Mary Lyon to Hannah White, February 26, 1834, in Lansing, *Mary Lyon Through Her Letters*, 129.

119. Letter from Mary Lyon, May 12, 1834, in Lansing, *Mary Lyon Through Her Letters*, 133–134.
120. Lyon's domestic system has been described in many places. See Green, *Mary Lyon and Mount Holyoke*, 114, 176–179; Allmendinger, Jr., "Mount Holyoke Students Encounter the Need for Life-Planning, 1837–1850," 33–34.
121. "Female Education," *The American Quarterly Observer* III (October 1834), 384–385.
122. See "Mount Holyoke Female Seminary," *American Annals of Education and Instruction* V (August 1835), 375–376 for an indication of the possible connection between the two.
123. "Domestic Seminary for Young Ladies," 499–503.
124. Farming was a common component of these systems, but some male colleges tried various forms of manufacturing, including barrel-making at the Marietta College and the Ohio University. See Rudolph, *The American College and University*, 217–218.
125. *First Circular of the Oberlin Collegiate Institute, March 8, 1834* (n.p.), 1.
126. Phelps, *The Female Student*, 130.
127. Sklar, "The Founding of Mount Holyoke College," 195–198; see also Mary Lyon's letter to Catharine Beecher, in which Lyon explains her thinking regarding fundraising, July 1, 1836, in Lansing, *Mary Lyon Through Her Letters*, 195–196.
128. Letter from Harriet Johnson to her sister, April 5, 1837, MHC, Ipswich Female Seminary, Records Series E.
129. T. H. Gallaudet, *An Address on Female Education, delivered, November 21st, 1827, at the Opening of the Edifice erected for the accommodation of the Hartford Female Seminary* (Hartford, CT: H. & F. J. Huntington, 1828); reprinted in *American Journal of Education* III (March 1828), 184–185.
130. Worcester, *An Address on Female Education*, 4.
131. Letter from Achsah Colburn to Rev. Mahan, October 24, 1836, Treasurer's Office File "B," Oberlin College Archives.
132. See the appendix for a list of schools examined.
133. Farnham says that "Southern elites insisted on 'accomplished' daughters." Christie Anne Farnham, *The Education of the Southern Belle: Higher Education and Student Socialization in the Antebellum South* (New York and London: New York University Press, 1994), 86.
134. Quote originally appeared in *New York Knickerbocker Magazine*; quoted in "Extremes in Female Education," *American Annals of Education and Instruction* (November 1835), 465.
135. Budden, *Thoughts on Domestic Education*, 278.
136. Marks, *Hints on Female Education*, 17. See also Lee, *The Contrast*, a novel in which a character speaks of music and painting as a "resource from ennui."
137. Charles Butler, *The American Lady* (Philadelphia: Hogan & Thompson, 1836), 57.
138. John Ludlow, *An Address Delivered at the Opening of the New Female Academy in Albany, May 12, 1834, By John Ludlow, D.D., President of the Board of*

*Trustees, with an Appendix* (Albany, NY: Packard & Van Benthuysen, printers, 1834), 8. Bradford Academy's catalog similarly states that a common defect in female education has been the "wish to render her a pretty and fashionable thing," rather than to make women useful. *Catalogue of the Officers and Members of Bradford Academy, Bradford, Massachusetts, for the year ending November 20, 1840* (Haverhill, MA: Essex Banner Press, 1840), 7. Similarly, the editors of the *American Journal of Education* critiqued reliance on a "showy and superficial course of study." See "Retrospect," *American Journal of Education* II (December 1827), 756.

139. "Extremes in Female Education," 465.

140. "Providence High School," *American Journal of Education* III (July 1828), 428.

141. *Annual Catalogue of the Officers, Teachers, and Pupils, of the Tuscaloosa Female Academy, for the Year 1832* (Tuscaloosa, AL: Expositor Office, printers, 1833), 10–12.

142. *Catalogue of the Officers, Teachers and Members of Day's Academy for Young Gentlemen, and Seminary for Young Ladies, from March 6, 1834, to January 14, 1835* (Boston: Beals and Greene, 1835), 16; *Catalogue of the Alabama Female Institute, Tuscaloosa, Ala. For the Year Ending 14th July, 1836* (Tuscaloosa, AL: Marmaduke J. Slade, 1836), 10. For two more examples, see: Clinton (North Carolina) Female Seminary (broadsheet, May 12, 1837, Duke Special Collections), which charged $100 for board, $32 for tuition, $50 for music, and $6 for use of the piano. At the Misses Watson School in Hartford, Connecticut, board and tuition cost $300 per year, music an extra $40, and use of piano another $10; see "Private Instruction" (broadsheet, March 1834, Henry Watson, Jr. papers, Duke Special Collections). See also Farnham, *The Education of the Southern Belle*, 86.

143. Nancy Beadie, "Female Students and Denominational Affiliation: Sources of Success and Variation among Nineteenth-Century Academies," *American Journal of Education* 107 (February 1999), 92–93.

144. Emerson, *Prospectus of the Female Seminary, at Wethersfield, CT*, 34.

145. Letter from Emma Willard to William Cogswell, January 10, 1842, Sophia Smith Collection.

146. *Catalogue of the Officers and Students of Amenia Seminary, 1841–42* (Poughkeepsie, NY: Jackson & Schram, 1842), 15.

147. Budden, *Thoughts on Domestic Education*, 71, 74–75.

148. Phelps, *The Female Student*, 29–30.

149. "Female Education" (November 1828), 649.

150. "Education of Females: Intellectual Instruction" (December 1827), 742.

151. Thomas Woody made a similar finding. He concluded that while 42% of schools offered plain needlework from 1749 to 1829, only 5% offered plain, and 12% offered fancy needlework from 1830 to 1871. Woody, *A History of Women's Education*, I, 418.

152. Susan Burrows Swan, *Plain and Fancy: American Women and Their Needlework, 1700–1850* (New York: Routledge, 1977), 204.

## 5  "PERFECTING OUR WHOLE NATURE"

1. Quoted in Ralph Emerson, *Life of Rev. Joseph Emerson, Pastor of the Third Congregational Church in Beverly, Ma. and subsequently Principal of a Female Seminary* (Boston: Crocker & Brewster, 1834), 421.

2. "Thoughts on the Education of Females," *American Journal of Education* I (June 1826), 349.

3. Charles Burroughs, *An Address on Female Education, Delivered in Portsmouth, New Hampshire, October 26, 1827* (Portsmouth: Childs and March, 1827), 6.

4. "Extremes in Female Education," originally published in *New York Knickerbocker*, reprinted in *American Annals of Education* IX (September 1835), 465.

5. "On Female Education," *American Journal of Education* III (May 1828), 162.

6. For discussion on the belief in women's intellectual inferiority, see Cynthia Kinnard, *Antifeminism in American Thought: An Annotated Bibliography* (Boston: G. K. Hall, 1986).

7. Burroughs, *An Address on Female Education*, 41.

8. *General View of the Plan of Education Pursued at the Adams Female Academy* (Exeter, NH: Nathaniel S. Adams, printer, 1831), 3.

9. Mrs. Townshend Stith, *Thoughts on Female Education* (Philadelphia: Clark & Raser, 1831), 30.

10. Quoted in "Neglect of Females," *American Annals of Education and Instruction* VI (November 1836), 523.

11. "Oberlin Collegiate Institute," *American Annals of Education* (October 1838), 477.

12. Stith, *Thoughts on Female Education*, 30.

13. Leonard Worcester, *An Address on Female Education. Delivered at Newark (N.J.) March 28, 1832* (Newark, NJ: n.p., 1832), 1–2. Worcester also writes, "The human mind is formed to be cultivated . . . [and] females are as capable as the other sex of being benefited by education."

14. "Female Education," *American Journal of Education* III (September 1828), 525.

15. Daniel Chandler, *An Address on Female Education, Delivered before the Demosthenian & Phi Kappa Societies, on the Day after Commencement, in the University of Georgia, by Daniel Chandler, Esq.* (Washington, GA: William A. Mercer, 1835), 12, 22.

16. S. S. Stocking, "An Address, Delivered Before the Young Ladies' Literary Society of the Wesleyan Academy, June 8, 1836" (Boston: David H. Ela, 1836), 5.

17. "Female Education," *The American Quarterly Observer* III (October 1834), 385.

18. Daniel Chandler, *An Address on Female Education*, 18–20.

19. *Third Annual Report of the Young Ladies' Association of the New-Hampton Female Seminary, for the Promotion of Literature and Missions; with the Constitution, etc. 1835–36* (Boston: John Putnam, 1837), 41–42. A similar expression of this view appeared in the *Ladies' Magazine* in 1834: "We will give the most charitable excuse for their long inattention to our wants in this respect, by supposing

they have concluded us so gifted by nature as to require little aid. . . ." Reprinted in "Female Education," *American Annals of Education and Instruction* IV (July 1834), 301.

20. Emma Willard to William Cogswell, January 10, 1842. Emma Willard Correspondence, Sophia Smith Collection.

21. Emma Willard, *An Address to the Public: Particularly to the Members of the Legislature of New York, Proposing a Plan for Improving Female Education* [1819] (Middlebury: Middlebury College, 1918). Also see Thomas Woody, *A History of Women's Education in the United States*, I (New York: The Science Press, 1929), 306–310; Nancy Beadie, "Emma Willard's Idea Put to the Test: The Consequences of State Support of Female Education in New York, 1819–67," *History of Education Quarterly* 33 (Winter 1993), 543–562.

22. Catharine Beecher, "Female Education," *American Journal of Education* II (May 1827), 268.

23. "Female Seminaries," *American Annals of Education and Instruction* VI (May 1836), 235–236.

24. See, for instance, Barbara Miller Solomon, *In the Company of Educated Women: A History of Women and Higher Education in America* (New Haven, CT and London: Yale University Press, 1985), 18–21; Anne Firor Scott, "The Ever-Widening Circle: The Diffusion of Feminist Values from the Troy Female Seminary, 1822–1872," *History of Education Quarterly* 19 (Spring 1979), 3–24; Helen Lefkowitz Horowitz, *Alma Mater: Design and Experience in the Women's Colleges from Their Nineteenth-Century Beginnings to the 1930s* (New York: Alfred A. Knopf, 1984), 11; David Tyack and Elisabeth Hansot, *Learning Together: A History of Coeducation in American Public Schools* (New Haven, CT and London: Yale University Press, 1990), 37.

25. Ralph Emerson, *Life of Rev. Joseph Emerson*, 248.

26. Joseph Emerson, *Prospectus of the Female Seminary, at Wethersfield, Ct. Comprising a General Prospectus, Course of Instruction, Maxims of Education, and Regulations of the Seminary* (Wethersfield, CT: A. Francis, 1826), 10.

27. *Catalogue of the Instructors and Students in the Female Classical Seminary, Brookfield, Mass.* (Brookfield, MA: E. & G. Merriam, 1826), 8; emphasis in original.

28. March 31, 1828, Albany Female Seminary Trustees' Minute Book, 1827–1849, vol. I, New York State Library.

29. *Catalogue of the Instructors and Pupils, in the New Haven Young Ladies' Institute, During its First Year* (New Haven, CT: n.p., 1830), 9.

30. *Fourth Annual Catalogue of the Teachers and Scholars, in the Young Ladies' High School, Boston. July, 1831* (Boston: W. W. Clapp, 1831), 11–12.

31. *Annual Catalogue of the Officers, Teachers, and Pupils, of the Tuscaloosa Female Academy for the Year 1832* (Tuscaloosa, AL: Expositor Office, 1833), 12, 15.

32. *Catalogue of the Officers and Students of Bradford Academy, Bradford, Massachusetts, October, 1839* (Haverhill, MA: E. H. Safford, 1839), 12.

33. *Mount Vernon Female Seminary* (Boston: n.p., 1836), 3.

34. "Female High-School of New-York," *American Journal of Education* I (January 1826), 59–60.

35. *Newark Institute for Young Ladies* (Newark: n.p., 1826), 7.
36. *Brooklyn Collegiate Institute for Young Ladies; Brooklyn Heights, Opposite the City of New York* (New York: n.p., 1830), 9.
37. *Catalogue of the Instructors and Pupils, in the New Haven Young Ladies' Institute,* 9–10.
38. "Greenfield High School for Young Ladies" (Greenfield, MA: broadsheet, 1840).
39. "Account of a Female School," *American Annals of Education and Instruction* II (April 1832), 213; emphasis in original.
40. Quoted in "Female Education," *American Annals of Education and Instruction* VII (October 1837), 447.
41. Day's statement, which subsequently became famous as the Yale Report of 1828, argued that classical languages were the best means to teach mental discipline, but allowed that other subjects, studied rigorously, might also accomplish the desired object. The Yale Report ([Jeremiah Day and James Kingsley], *Reports on the Course of Instruction in Yale College; by a Committee of the Corporation and the Academical Faculty*) was published in the *American Journal of Science and Arts* 15 (January 1829), 297–351. There are numerous discussions of the Yale Report. See Roger L. Geiger, "The Historical Matrix of American Higher Education," *History of Higher Education Annual* 12 (1992), 13; Stanley M. Guralnik, *Science and the Ante-Bellum American College* (Philadelphia: American Philosophical Society), 1975, 28–33.
42. Elias Marks, *Hints on Female Education, with an outline of an Institution for the Education of Females, Termed the So. Ca. Female Institute; under the direction of Dr. Elis Marks* (Columbia, SC: David W. Sims, 1828), 19.
43. *General View of the Plan of Education,* 3.
44. Ralph Emerson, *Life of the Rev. Joseph Emerson* quoted in "Life of Emerson," *American Annals of Education and Instruction* IV (August 1834), 344; emphasis in original.
45. "Female Education," *American Journal of Education* III (September 1828), 521.
46. Ibid.
47. Quoted in "Ladies' Association for the Education of Female Teachers," *American Annals of Education and Instruction* V (January 1835), 83.
48. Kim Tolley argues that before 1840, "textbooks used in female schools generally emphasized a conceptual, rather than a mathematical, understanding of scientific principles," but that after 1840, the more advanced science textbooks used in female schools were comparable to those offered in male schools. She makes a similar argument for physics. See Kim Tolley, *The Science Education of American Girls: A Historical Perspective* (New York and London: RoutledgeFalmer, 2003), 8; see especially chs. 2–4.
49. "Thoughts on the Education of Females" (June 1826), 351.
50. "Providence High School," reprinted in *American Journal of Education* III (July 1828), 428; italics in original.
51. *Fourth Annual Catalogue of the Teachers and Scholars, in the Young Ladies' High School,* 13.
52. *Third Annual Report of the Young Ladies' Association of the New-Hampton Female Seminary,* 45–46.

53. *General View of the Plan of Education*, 4.

54. "Female Education," *The American Quarterly Observer* III (October 1834), 385; emphasis in original.

55. Maria Budden, *Thoughts on Domestic Education; the Result of Experience. By a Mother* (London: Charles Knight, 1826), 59, 64–65.

56. "Account of a Female School" (April 1832), 213; emphasis in original.

57. "Education of Females," *American Journal of Education* II (December 1827), 739.

58. *Young Ladies' Institute Circular (1828)*, reprinted in *Catalogue of the Instructors and Pupils, in the New Haven Young Ladies' Institute*, 6.

59. *An Appeal to Parents for Female Education on Christian Principles; with a Prospectus of St. Mary's Hall, Green Bank, Burlington, New Jersey* (Burlington, NJ: J.L. Powell Missionary Press, 1837), 10, 11, 17, 24; italics and emphasis of capitalization in original.

60. *Ballston Spa Female Academy* (Albany, NY: Packard & Van Benthuysen, 1824), 3–8.

61. John T. Irving, *Address Delivered on the Opening of the New-York High-School for Females, January 31, 1826* (New York: William A. Mercein, 1826), 20–21, 24.

62. *Annual Catalogue of the Instructers* [sic] *and Pupils in the Newburgh (Late Mount Pleasant) Female Seminary. Situated in Newburgh, Orange Co. N.Y. for the Year preceding January 1st, 1837* (Newburgh, NY: J. D. Spalding, 1837), 8.

63. Joseph Emerson, *Prospectus of the Female Seminary, at Wethersfield, Ct.*, 59.

64. Joseph Emerson to Dr. Humphrey, January 24, 1827, Joseph Emerson Papers, Mount Holyoke College Archives.

65. For lists of schools surveyed, see the appendix in this book. Percentages of schools offering the following subjects in the 1830s were

|           | South | West | Mid-Atlantic | New England |
|-----------|-------|------|--------------|-------------|
| Algebra   | 50    | 82   | 53           | 80          |
| Geometry  | 63    | 64   | 65           | 92          |
| Chemistry | 88    | 82   | 71           | 72          |
| Botany    | 75    | 82   | 76           | 64          |
| Astronomy | 56    | 91   | 71           | 64          |

66. For a list of schools surveyed, see the appendix.

67. "Education of Females," *American Journal of Education* I (July 1826), 350.

68. Beecher, "Female Education" (May 1827), 268.

69. "Female Education" (October 1837), 447–449; italics in original.

70. Marks, *Hints on Female Education*, 31–34.

71. Leonard Worcester, *An Address on Female Education*, 6.

72. *Fourth Annual Catalogue of the Teachers and Scholars, in the Young Ladies' High School*, 14.

73. *General View of the Plan of Education*, 8–9.

74. *Catalogue of the Instructors and Pupils of the Young Ladies' Institute, New Haven, Conn., for the Year Ending April, 1839* (New Haven, CT: Hitchcock & Stafford, 1839), 10; emphasis in original.

75. *Roxbury Female School* (Roxbury: n.p., 1830?), 11; emphasis in original.
76. *Fifth Annual Catalogue of the Teachers & Scholars in the Gothic Seminary, Northampton, Mass. September, 1840* (Northampton: n.p., 1840), 9.
77. For example, see "Education of Females," *American Journal of Education* II (December 1827), 741. For a brief discussion on the popularity of Pestalozzi among antebellum educational reformers, see Woody, *A History of Women's Education*, I, 317.
78. "Annual Report Presented to the Trustees of the Elizabeth Female Academy, by Mrs. C. M. Thayer, Governess," reprinted in *American Journal of Education* II (October 1827), 633–634; emphasis in original.
79. Scott, "The Ever-Widening Circle," 8.
80. Introductory Lecture at Byfield Seminary, May 1821, notes taken by Mary Lyon, Joseph Emerson Papers, Mount Holyoke College Archives; emphasis in original.
81. Lectures on Intellectual Philosophy, Ipswich Female Seminary, 1834, Mount Holyoke College Archives.
82. James M. Garnett, *Lectures on Female Education, Comprising the First and Second Series of a Course Delivered to Mrs. Garnett's Pupils, At Elm-Wood, Essex County, Virginia. By James M. Garnett. To Which is Annexed, The Gossip's Manual* (Richmond, VA: Thomas W. White, 1825), 86–87, 92, 255–256; emphasis in original.
83. "Suggestions useful in teaching," handwritten notes, possibly from classes led by either Mary Lyon or Zilpah Grant, ca. 1835, Ipswich Female Academy, Mount Holyoke College Archives.
84. T. H. Gallaudet, *An Address on Female Education, delivered, November 21st, 1827, at the Opening of the Edifice erected for the accommodation of the Hartford Female Seminary*, (Hartford, CT: H. & F. J. Huntington, 1828), reprinted in *American Journal of Education* III (March 1828), 179.
85. "Some of the Characteristics of Ipswich Female Academy, Communicated by the Senior Class," *Catalogue of the Officers and Members of the Ipswich Female Academy* (Ipswich, MA: John Harris, Jr., 1829), 7.
86. *Catalogue of the Mount Vernon Female School. Containing the Names of the Trustees, Teachers and Pupils, in January, 1831* (Boston: T. R. Marvin, 1831), 10.
87. *Arcade Ladies' Institute, Providence R.I.* (Providence, RI: H. H. Brown, 1835), 9.
88. "Private Instruction," broadsheet (Hartford, CT: n.p., 1834), Duke Special Collections.
89. *Catalogue of the Alabama Female Institute, Tuscaloosa, Ala. For the Year Ending 14th July, 1836* (Tuscaloosa, AL: Marmaduke J. Slade, 1836), 11; emphasis in original.
90. Mary Ann Adams, Ellen F. Griswold, and Elizabeth S. Peck to John Keep and William Dawes, July 10, 1839, Keep Papers, Robert S. Fletcher Files, Oberlin College Archives; emphasis in original.
91. "Female Education," *American Journal of Education* III (September 1828), 525; emphasis in original.
92. "Popular Education," *North American Review* 36 ( January 1833), 82–83.
93. "Emulation," *American Annals of Education and Instruction* VI (March 1836), 109.

94. "Education of Females," *American Journal of Education* II (November 1827), 550–551. See also William J. Reese, *The Origins of the American High School* (New Haven, CT and London: Yale University Press, 1995), 46.

95. "Boston High School for Girls," *American Journal of Education* II (March 1827), 186.

96. "Motives to Study in the Ipswich Female Seminary," *American Annals of Education and Instruction* III (February 1833), 75, 79.

97. Andrea Kerr, *Lucy Stone: Speaking Out for Equality* (New Brunswick, NJ: Rutgers University Press, 1992), ch. 2.

98. "Introductory Lecture at the Seminary, May 1821," notes taken by Mary Lyon, Joseph Emerson Papers, Folder 2, Manuscript lectures and other writings, 1804–1831, Mount Holyoke College Archives.

99. Garnett, *Lectures on Female Education*, 58, 74, 266.

100. Budden, *Thoughts on Domestic Education*, 219–220.

101. Stith, *Thoughts on Female Education*, 13–15.

102. *Catalogue of the Mount Vernon Female School*, 10, 11.

103. Chandler, *An Address on Female Education*, 12.

104. E. Becket, "Education," Compositions by students at Adams Female Academy, November, 1824, Mount Holyoke College Archives.

105. Maria Cowles to Henry Cowles, November 25, 1829, Ipswich Female Seminary, Series E, Folder 3, Mount Holyoke College Archives.

106. Mary Lyon to Zilpah Grant, December 26, 1825, in Marion Lansing, ed., *Mary Lyon Through Her Letters* (Boston: Books, Inc., 1937), 61.

107. Almira Hart Lincoln Phelps, *The Female Student; Or, Lectures to Young Ladies on Female Education* (New York: Leavitt, Lord & Co., 1836), 53.

108. S. F. W., "Female Education: Extract from A Letter to the Editor of *Ladies' Magazine*," reprinted in *American Annals of Education and Instruction* IV ( July 1834), 300; emphasis in original.

109. *Second Annual Report of the Young Ladies' Association of the New-Hampton Female Seminary, for the Promotion of Literature and Missions; with the Constitution, etc. 1834–5* (Boston: Freeman and Bolles, 1836), 27.

110. Mary Ann Adams, Ellen F. Griswold, and Elizabeth S. Peck to John Keep and William Dawes, July 10, 1839, Keep Papers, Robert S. Fletcher Files, Box 7, Oberlin College Archives.

111. *An Account of the High School for Girls, Boston: with A Catalogue of the Scholars. February, 1826* (Boston: Thomas B. Wait and Son, 1826), 7, 12–13.

112. Martha Whiting, journal entry, June 1, 1835, in Catharine Badger, *The Teacher's Last Lesson: A Memoir of Martha Whiting, Late of the Charlestown Female Seminary. Consisting Chiefly of Extracts from Her Journal, Interspersed with Reminiscences and Suggestive Reflections* (Boston: Gould and Lincoln, 1855), 132; emphasis in original.

113. Eliza Mira Lenoir to Julia Pickens, October 7, 1828, Chiliab Howe Papers, Southern Historical Collection, UNC.

114. Harriet Cary to Elizabeth Wainwright, March 20, 1816, Peter Wainwright, Jr. Papers, Duke Special Collections.

115. S. F. W., "Female Education," 300; emphasis in original.

116. In addition to quotes already cited in this chapter, see "Thoughts on the Education of Females" (July 1826), 349, in which an anonymous essayist writes of women, "Science allures them to her temple, and virtue commands them to dedicate to her altar, that influence which they derive from the courtesy of refined society."

117. "Course of Education in the New-York High-School," *American Journal of Education* I (January 1826), 23.

118. "Address on Associations to Promote Education," *American Annals of Education and Instruction* V (July 1835), 317. For examples of school catalogs that state this philosophy, see: *Catalogue of the Greenfield High School for Young Ladies* (Greenfield, MA: Phelps and Ingersoll, 1830), 6 ("the education we contemplate is not limited to instruction. Our plan embraces physical and moral, as well as mental culture"); John W. Scott, *An Address on Female Education, Delivered At the Close of the Summer Session for 1840, of the Steubenville Female Seminary, in Presence of its Pupils and Patrons* (Steubenville, OH: n.p., 1840), 6 (in addition to mental culture, education should promote the "growth, health and symmetrical development of all the powers and capacities of the body"); *Circular of the Smithville Academy* (Providence, RI: n.p., 1838), 1 ("to thoroughly prepare . . . young persons of both sexes . . . requires an Institution that will educate the whole *Physical, Intellectual and Moral being of the individual student*" [emphasis in original]).

119. James Clark, "A Treatise on Pulmonary Consumption," quoted in "Errors in Physical Education," *American Annals of Education and Instruction* VI (March 1836), 106.

120. Daniel Mayes, *An Address Delivered on the First Anniversary of Van Doren's Collegiate Institute for Young Ladies, in the City of Lexington, Ky. On the Last Thursday of July, 1832* (Lexington, KY: Finnell & Herndon, 1832), 3.

121. "Thoughts on the Education of Females" (July 1826), 351.

122. John Andrews, "Physical Education of Females," *American Annals of Education and Instruction* VI (February 1836), 85.

123. "Review," *American Journal of Education* II (July 1827), 428.

124. Quoted in "Female Education," *American Annals of Education and Instruction* VI (March 1836), 101.

125. Much of this discussion centers around the publication of Dr. Edward Clark's *Sex in Education, or a Fair Chance For the Girls* (Boston: J. R. Osgood) in 1873, and his *The Building of a Brain* (Boston: J. R. Osgood) in 1874. For historians' treatment of this issue, see Lynn D. Gordon, *Gender and Higher Education in the Progressive Era* (New Haven, CT and London: Yale University Press, 1990), 18–21; Solomon, *In the Company of Educated Women*, 56–57.

126. See, e.g., "Appropriate Exercise," *The Journal of Health* I (September 23, 1829), 22; "Variety in Exercise," *The Journal of Health* I (April 28, 1830), 243–244; "Benefits of Exercise," *The Journal of Health* 2 (November 30, 1830), 91–92; "Suggestions to Parents: Physical Education," *The American Journal of Education* 1 (April 1826), 235–239.

127. "Insanity from Excessive Study," *American Annals of Education and Instruction* III (September 1833), 425.

128. "Errors in Physical Education," *American Annals of Education and Instruction* VI (March 1836), 107.
129. Dr. John Bell, "Physical Education of Girls," *Journal of Health* I (September 9, 1829), 15.
130. "Review of Charles Londe, *Medical Gymnastics*," *American Journal of Education* I (April 1826), 235–239.
131. "Buffalo High School," *American Annals of Education and Instruction* 3 (April 1828), 233–235.
132. "Calisthenics," *Journal of Health* 2 (February 23, 1831), 190.
133. "Thoughts on the Education of Females" (June 1826), 352. See also Jan Todd, *Physical Culture and the Body Beautiful: Purposive Exercise In the Lives of American Women, 1800–1870* (Macon, GA: Mercer University Press, 1998).
134. "Gymnastic Exercise for Females," *American Journal of Education* I (November 1826), 698, reprinted from the *Medical Intelligencer*.
135. "Western Female Institute," *American Annals of Education and Instruction* III (August 1833), 380–381.
136. *A Course of Calisthenics for Young Ladies in Schools and Families with Some Remarks on Physical Education* (Hartford, CT: H. and F. J. Huntington, 1831), 36–37.
137. Sophia Peck to F. and Eliza Peck, January 23, 1837, Henry Watson, Jr. Papers, Duke Special Collections.
138. "Education of Females," *American Journal of Education* II (July 1827), 425.
139. *Outline and Catalogue of the Steubenville Female Seminary for the Year Ending in October, 1840* (Steubenville, OH: n.p., 1840), 3.
140. *Arcade Ladies' Institute*, 7.
141. *Catalogue of the Instructors and Pupils, in the New Haven Young Ladies' Institute*, 9.

## 6   POSSIBILITIES AND LIMITATIONS

1. Carl F. Kaestle, *Pillars of the Republic: Common Schools and American Society, 1780–1860*, (New York: Hill and Wang, 1983), 118; Thomas Dublin, ed., *Farm to Factory: Women's Letters, 1830–1860* (New York: Columbia University Press, 1981), 21–22.
2. Mary Lyon to Hannah White, February 26, 1834, in Marion Lansing, ed., *Mary Lyon Through Her Letters* (Boston: Books, Inc., 1937), 129.
3. *First Circular of the Oberlin Collegiate Institute, March 8, 1834* (Oberlin: n.p., 1834), 1.
4. For examples, see *Catalogue of the Officers and Students of Bradford Academy, Bradford, Massachusetts, October, 1839* (Haverhill, MA: E. H. Safford, 1839), 12; reports of the Young Ladies' Association of the New-Hampton Female Seminary; and the Ipswich Society for the Education of Females.
5. Nina Baym, "Women and the Republic: Emma Willard's Rhetoric of History," *American Quarterly* 43 (March 1991), 6–7.

6. Emma Willard, *An Address to the Public: Particularly to the Members of the Legislature of New York, Proposing a Plan for Improving Female Education* [1819] (Middlebury, VT: Middlebury College, 1918); Baym, "Women and the Republic," 7–8.

7. Ellen N. Lawson and Marlene Merrill, "The Antebellum 'Talented Thousandth': Black College Students at Oberlin Before the Civil War," *Journal of Negro Education* 52 (Spring 1983), 154.

8. Madelyn Holmes and Beverly J. Weiss, *Lives of Women Public Schoolteachers: Scenes from American Educational History* (New York and London: Garland Publishing Inc., 1995), 63, 192.

9. For one account of the incident, see Shirley J. Yee, *Black Women Abolitionists: A Study in Activism, 1828–1860* (Knoxville: University of Tennessee Press, 1992), 50–51. Also see Bonnie Handler, "Prudence Crandall and Her School for Young Ladies and Little Misses of Color," *Vitae Scholastica* 5 (Winter 1986), 199–210; Philip S. Foner and Josephine F. Pacheco, *Three Who Dared: Prudence Crandall, Margaret Douglass, Myrtilla Miner: Champions of Antebellum Black Education* (Westport, CT: Greenwood Press, 1984).

10. See *Freedom's Journal* (August 24, 1827); Sarah Mapps Douglass, "Address," *The Liberator* (July 21, 1832); "Constitution, Female Literary Association," *The Liberator* (December 3, 1831); Dorothy B. Porter, "The Organized Educational Activities of Negro Literary Societies, 1828–1846," *Journal of Negro Education* 5 (October 1936), 555–576; Julie Winch, " 'You Have Talents—Only Cultivate Them': Philadelphia's Black Female Literary Societies and the Abolitionist Crusade," in *The Abolitionist Sisterhood: Women's Political Culture in Antebellum America*, Jean Fagan Yellin and John C. Van Home, eds. (Ithaca, NY: Cornell University Press, 1994), 101–118; Yee, *Black Women Abolitionists*; James Oliver Horton and Lois E. Horton, *In Hope of Liberty: Culture, Community, and Protest among Northern Free Blacks, 1700–1860* (New York: Oxford University Press, 1997). Also see Mary Kelley, " 'A More Glorious Revolution': Women's Antebellum Reading Circles and the Pursuit of Public Influence," *New England Quarterly* 76 (June 2003), 163–196.

11. Maria W. Stewart, "Religion and the Pure Principles of Morality, The Sure Foundation on Which We Must Build," reprinted in *Maria W. Stewart, America's First Black Woman Political Writer: Essays and Speeches*, Marilyn Richardson, ed. (Bloomington: Indiana University Press, 1987), 38.

12. Fanny Jackson Coppin, principal of the Institute for Colored Youth, went to the Rhode Island State Normal School in 1859, prior to attending Oberlin. Charlotte Forten Grimke, who became well known for the journals she kept while she taught recently freed people as part of the Port Royal Project in South Carolina, attended the Normal School at Salem (Massachusetts) in 1854. Sarah Smith, who married Henry Highland Garnet, attended several normal schools in the New York City area in the 1850s, although we do not know which ones. Myrtilla Miner's School for Colored Girls in Washington, DC opened in 1851, training women to be teachers. See Linda M. Perkins, "Heed Life's Demands: The Educational Philosophy of Fanny Jackson Coppin," *Journal of Negro Education* 51 (Summer 1982), 181; Janice Sumler-Edmond, "Charlotte L. Forten Grimke," in *Black*

*Women in America: An Historical Encyclopedia*, Darlene Clark Hine et al., eds. (Bloomington: Indiana University Press, 1993), 505; "Sarah S. T. Garnet," in *Black Women in America*, 479; Dorothy Sterling, *We Are Your Sisters: Black Women in the Nineteenth Century* (New York and London: W. W. Norton & Co., 1984), 190.

13. For example, Arcade Ladies' Institute (Rhode Island), Brooklyn Collegiate Institute for Young Ladies, Greenfield High School for Young Ladies (Massachusetts), Newark Institute for Young Ladies, New Haven Young Ladies' Institute, Van Doren's Collegiate Institute for Young Ladies, Western Collegiate Institute for Young Ladies (Pennsylvania), the Young Ladies' High School (Boston), the Young Ladies' Seminary (Keene, New Hampshire). Most seminaries, however, used the terms "female" or "women."

14. Political scientist Pauline Schloesser describes the process of the construction of white women as a "racialized sex group that lost consciousness of itself as bounded by race and class." Pauline Schloesser, *The Fair Sex: White Women and Racial Patriarchy in the Early American Republic* (New York and London: New York University Press, 2002), 53.

15. Christine Stansell, *City of Women: Sex and Class in New York, 1789–1860* (Urbana and Chicago: University of Illinois Press, 1987), 68. For other work on class formation in this era, see Stuart M. Blumin, Robert Fogel, and Stephan Thernstrom, eds., *The Emergence of the Middle Class: Social Experience in the American City, 1760–1900* (New York: Cambridge University Press, 1989); Linda Young, *Middle Class Culture in the Nineteenth Century: America, Australia and Britain* (New York: Palgrave Macmillan, 2003); Jonathan Daniel Wells, *The Origins of the Southern Middle Class, 1800–1861* (Chapel Hill: The University of North Carolina Press, 2004); Heidi L. Nichols, *The Fashioning of Middle-Class America: Sartains Union Magazine of Literature and Art and Antebellum Culture* (New York: Peter Lang, 2004); T. Walter Herbert, *Dearest Beloved: The Hawthornes and the Making of the Middle-Class Family* (Berkeley: University of California Press, 1993); Burton S. Bledstein and Robert D. Johnston, eds., *The Middling Sorts: Explorations in the History of the American Middle Class* (New York and London: Routledge Press, 2000).

16. New York City Tract Society reports of 1837, 1838, and 1839, quoted in Stansell, *City of Women*, 63.

17. See Kaestle, *Pillars of the Republic*, especially ch. 6; William J. Reese, *The Origins of the American High School* (New Haven, CT and London: Yale University Press, 1995), especially ch. 3.

18. Michael Chevalier, *Society, Manners and Politics in the United States: Being a Series of Letters on North America* (Boston: Weeks, Jordan & Company, 1839), 137; Anthony Trollope, *North America* [1861] (New York: St. Martin's, 1986), 249; Amal Amireh, *The Factory Girl and the Seamstress: Imagining Gender and Class in Nineteenth Century American Fiction* (New York and London: Garland, 2000), 7. Also see Gerda Lerner, "The Lady and the Mill Girl: Changes in the Status of Women in the Age of Jackson," *Midcontinent American Studies Journal* 10 (Winter 1969), 5–15.

19. *Catalogue of the Instructors and Pupils, in the New Haven Young Ladies' Institute, During Its First Year* (New Haven, CT: n.p., 1830), 7–8.

20. For discussions of respectability, see Richard L. Bushman, *The Refinement of America: Persons, Houses, Cities* (New York: Random House, 1992); Blumin et al., *The Emergence of the Middle Class*; Daniel A. Cohen, "The Respectability of Rebecca Reed: Genteel Womanhood and Sectarian Conflict in Antebellum America," *Journal of the Early Republic* 16 (Fall 1996), 419–461.

21. For examples, see *Newark Institute for Young Ladies* (Newark, NJ: n.p., 1826), 7.

22. Elias Marks, *Hints on Female Education, with an Outline of an Institution for the Education of Females, Termed the So. Ca. Female Institute; under the direction of Dr. Elias Marks* (Columbia, SC: David W. Sims, 1828), 42.

23. *Catalogue of the Alabama Female Institute, Tuscaloosa, Ala. For the Year Ending 14th July, 1836* (Tuscaloosa, AL: Marmaduke J. Slade, 1836), 11.

24. Baynard R. Hall, *An Address Delivered to the Young Ladies of the Spring-Villa Seminary, at Bordentown, N.J. At the Distribution of the Annual Medal and Premiums: on the Evening of the 29th of August, 1839* (Burlington NJ: Powell & George, 1839), 8.

25. *Arcade Ladies' Institute, Providence, R.I.* (Providence, RI: n.p., 1834?), 7.

26. *Ballston Spa Female Seminary* (Albany, NY: Packard & Van Benthuysen, 1824), 5.

27. *Catalogue of the Greenfield High School for Young Ladies, for the Year 1836–37* (Greenfield, MA: Phelps & Intersoll, 1837), 9; John W. Scott, *An Address on Female Education, Delivered at the Close of the Summer Session for 1840, of the Steubenville Female Seminary, in Presence of its Pupils and Patrons* (Steubenville, OH: n.p., 1840), 6.

28. *Prospectus of the Lexington Female Academy* (Lexington, KY: n.p., 1821), 2.

29. *Roxbury Female School* (Boston? n.p., 1830?), 11, 13; emphasis in original.

30. *Brooklyn Collegiate Institute for Young Ladies; Brooklyn Heights, Opposite the City of New York* (New York: n.p., 1830), 8; emphasis in original.

31. *Second Annual Report of the Young Ladies' Association of the New-Hampton Female Seminary, for the Promotion of Literature and Missions; with the Constitution, etc. 1839–40* (Boston: Freeman and Bolles, 1836), 29.

32. *Arcade Ladies' Institute*, 8.

33. *Third Annual Report of the Young Ladies' Association of the New-Hampton Female Seminary, for the Promotion of Literature and Missions; with the Constitution, etc. 1835–36* (Boston: John Putnam, 1837), 22.

34. *Catalogue of the Young Ladies' Seminary, in Keene, N.H. for the Year Ending October, 1832* (Keene, NH: J. & J. W. Prentiss, 1832), 13.

35. Abigail Mott, *Observations on the Importance of Female Education, and Maternal Instruction, with their Beneficial Influence on Society. By a Mother* (New York: Mahlon Day, 1825), 14.

36. *Mount Vernon Female Seminary* (Boston: n.p., 1836), 11.

37. *Catalogue and Circular of the LeRoy Female Seminary* (Rochester, NY: David Hoyt, 1840), 11; *Catalogue of the Officers and Students of the Townsend Female Seminary, for the Year Ending March, 1839* (Boston: John Putnam, 1839), 3.

38. *Catalogue of the Greenfield High School for Young Ladies, for the Year 1836–37*, 8.

39. Mrs. Townshend Stith, *Thoughts on Female Education* (Philadelphia: Clark & Raser, 1831), 13.

40. Almira Hart Lincoln Phelps, *The Female Student; Or, Lectures to Young Ladies on Female Education* (New York: Leavitt, Lord & Co., 1836), 29. See also James M. Garnett, *Lectures on Female Education, Comprising the First and Second Series of a Course Delivered to Mrs. Garnett's Pupils, At Elm-Wood, Essex County, Virginia. By James M. Garnett. To Which is Annexed, The Gossip's Manual* (Richmond, VA: Thomas W. White, 1825), 58–62.

41. Some coeducational academies specified that they were for "young gentlemen" and "young ladies." See, for instance, *Catalogue of the Officers, Teachers and Members of Day's Academy for Young Gentlemen, and Seminary for Young Ladies from March 6, 1834, to January 14, 1835* (Boston: Beals & Greene, 1835).

42. William J. Adams, *Circular* (New York: n.p., 1829), 2, 4, 5.

43. James Abercrombie, *A Charge Delivered, after a Public Examination, On Friday, July 27, 1804, to the Senior Class of the Philadelphia Academy, upon Their having Completed the Course of Study Prescribed by that Institution* (Philadelphia: H. Maxwell, 1804), 8.

44. Catharine M. Sedgwick, *Means and Ends, or Self-Training* (Boston: March, Capen, Lyon & Webb, 1839), 267; emphasis in original.

45. Thomas Gisborne, *An Enquiry into the Duties of the Female Sex* (London: T. Cadell and W. Davies, 1798), 19; Charles Butler, *The American Lady* (Philadelphia: Hogan & Thompson, 1836), 19. Butler plagiarized so completely that even the pagination remained the same for much of the two books. However, in the passage quoted here, Butler added the hostile phrase "the supposed lords of creation" to his description of women's views of men.

46. Gisborne, *An Enquiry into the Duties of the Female Sex*, 107.

47. Butler, *The American Lady*, 185; emphasis in original.

48. Catherine A. Brekus, *Strangers & Pilgrims: Female Preaching in America, 1740–1845* (Chapel Hill: University of North Carolina Press, 1998), 136.

49. Henry Barnard, ed., *Memoirs of Teachers, Educators, and Promoters and Benefactors of Education, Literature, and Science. Reprinted from the American Journal of Education* (New York: F. C. Brownell, 1861), 133.

50. Nancy Beadie, "Emma Willard's Idea Put to the Test: The Consequences of State Support of Female Education in New York, 1819–67," *History of Education Quarterly* 33 (Winter 1993), 543–545.

51. Quoted in Thomas Woody, *A History of Women's Education in the United States*, I (New York: The Science Press, 1929), 308.

52. For quotation, see Anne Firor Scott, "The Ever-Widening Circle: The Diffusion of Feminist Values from the Troy Female Seminary, 1822–1872," *History of Education Quarterly* 19 (Spring 1979), 7. For curriculum, see ibid., 7; Lucy Forsyth Townsend, "Emma Willard: Eclipse or Reemergence?" *Journal of the Midwest History of Education Society* 18 (1990), 289–290.

53. Quoted in Scott, "The Ever-Widening Circle," 22, fn. 8.

54. Townsend, "Emma Willard," 290.

55. Scott, "The Ever-Widening Circle," 11–12.

56. Baym, "Women and the Republic," 6.

57. Ibid., 5–6.

58. Scott, "The Ever-Widening Circle," 11–12.

59. Lucy F. Townsend and Barbara Wiley, "Divorce and Domestic Education: The Case of Emma Willard," Unpublished paper, 10, 19–20; Townsend, "Emma Willard," 285.

60. The best works on Beecher are Kathryn Kish Sklar, *Catharine Beecher: A Study in American Domesticity* (New Haven, CT: Yale University Press, 1973), and Jeanne Boydston, Mary Kelley, and Anne Margolis, *The Limits of Sisterhood: The Beecher Sisters on Women's Rights and Woman's Sphere* (Chapel Hill and London: University of North Carolina Press, 1988).

61. Catharine Beecher, *The True Remedy for the Wrongs of Women* (Boston: Phillips, Sampson, 1851), quoted in Boydston, Kelley, and Margolis, *The Limits of Sisterhood* 139; emphasis in original.

62. Quoted in Elizabeth Alden Green, *Mary Lyon and Mount Holyoke: Opening the Gates* (Hanover, NH: University Press of New England, 1979), 160.

63. Mary Lyon to Catharine Beecher, July 1, 1836, in *Mary Lyon Through Her Letters*, Marion Lansing, ed. (Boston: Books, Inc., 1937), 199.

64. Joseph A. Conforti, *Jonathan Edwards, Religious Tradition, and American Culture* (Chapel Hill and London: University of North Carolina Press, 1995).

65. Mary Lyon to Zilpah Grant, February 4, 1833, in *Mary Lyon Through Her Letters*, Lansing, ed., 110; emphasis in original.

66. Mary Lyon to Catharine Beecher, July 1, 1836, 199.

67. See chapter 1 in this book for a discussion of this.

68. The "cult of true womanhood" was first defined by Barbara Welter in her "The Cult of True Womanhood," *American Quarterly* 18 (Summer 1966), 151–174.

69. Frances B. Cogan, *All-American Girl: The Ideal of Real Womanhood in Mid-Nineteenth-Century America* (Athens: University of Georgia Press, 1989).

70. Mary Kelley, "The Sentimentalists: Promise and Betrayal in the Home," *Signs* 4 (Spring 1979); Mary Kelley, *Private Woman Public Stage: Literary Domesticity in Nineteenth-Century America* (New York: Oxford University Press, 1984).

71. Laura McCall, " 'The Reign of Brute Force Is Now Over': A Content Analysis of *Godey's Lady's Book, 1830–1860,"* *Journal of the Early Republic* 9 (Summer 1989), 235; emphasis in original.

72. Laura McCall, " 'Shall I Fetter Her Will?': Literary Americans Confront Feminine Submission, 1820–1860," *Journal of the Early Republic* 21 (Spring 2001), 97–99.

73. Rev. H. Hutchins, quoted in Catharine Naomi Badger, *The Teacher's Last Lesson: A Memoir of Martha Whiting, Late of the Charlestown Female Seminary. Consisting Chiefly of Extracts from Her Journal, Interspersed with Reminiscences and Suggestive Reflections* (Boston: Gould and Lincoln, 1855), 263.

74. Brekus, *Strangers & Pilgrims*, 153.

75. Mott, *Observations on the Importance of Female Education*, 49, 13.

76. *General View of the Plan of Education Pursued at the Adams Female Academy* (Exeter, NH: Nathaniel S. Adams, printer, 1831), 5; *Catalogue of the Greenfield High School for Young Ladies, for the Year 1836–37*, 9; *Catalogue of the Officers and Members of the Utica Female Academy* (Utica, NY: Bennett, Backus, & Hawley, 1840), 17.

77. Marcus Cicero Stephens to Mary Ann Primrose, November 7, 1841, Stephens Papers, Southern Historical Collection.
78. See, e.g., Scott, "The Ever-Widening Circle," 8, 23, fn. 12.
79. Scott, *An Address on Female Education*, 10.
80. S. S. Stocking, *An Address, Delivered Before the Young Ladies' Literary Society of the Wesleyan Academy, June 8, 1836* (Boston: David H. Ela, 1836), 22; emphasis in original.
81. *A Catalogue of the Abbot Female Seminary, Andover, Mass* (Andover, MA: Gould, Newman and Saxton, 1840), 11; emphasis in original; *Catalogue of the Alabama Female Institute, Tuscaloosa Ala. For the Year Ending 14th July, 1836*, 11. For more examples of self-government in female seminaries, see *Greenfield High School for Young Ladies* (Greenfield, MA: n.p., 1840), 1. "Our authority is sustained by reason": *Outline and Catalogue of the Steubenville Female Seminary for the Year Ending in October, 1840* (Steubenville, OH: n.p., 1840), 6.
82. *Catalogue of the Trustees, Instructors and Pupils in the Uxbridge Female Seminary, at Uxbridge, Mass* (Providence, RI: H. H. Brown, 1838), 9.
83. *Western Collegiate Institute* (Pittsburgh: n.p., 1837), 1.
84. *Roxbury Female School*, 13.
85. "Declaration of Sentiments," reprinted in *Women's America: Refocusing the Past*, 2nd ed., Linda K. Kerber and Jane DeHart-Mathews, eds. (New York and Oxford: Oxford University Press, 1987), 472.
86. Elizabeth Cady Stanton, "Address to the Legislature of New York on Women's Rights," February 14, 1854, in *The Elizabeth Cady Stanton-Susan B. Anthony Reader: Correspondence, Writings, Speeches*, Ellen Carol DuBois, ed. (Boston: Northeastern University Press, 1992), 45.
87. Some activists had already pitted "women's" issues (i.e., white middle-class women) against the issue of abolition as early as 1840. Abby Kelley and Frederick Douglass, for instance, were on opposing sides of this issue that effectively caused the "first grand division" in the antislavery movement. See Dorothy Sterling, *Ahead of Her Time: Abby Kelley and the Politics of Antislavery* (New York and London: W. W. Norton & Co., 1991), ch. 14. However, the issue became much more heated and divisive after the Civil War. There are many accounts of the racism within the women's rights campaigns of the post–Civil War era. See Ellen Carol DuBois, *Feminism and Suffrage: The Emergence of an Independent Women's Movement in America, 1848–1869* (Ithaca, NY and London: Cornell University Press, 1978); Andrea Moore Kerr, *Lucy Stone: Speaking Out for Equality* (New Brunswick, NJ: Rutgers University Press, 1992), ch. 8.
88. See Reports from Young Ladies' Association of the New-Hampton Female Seminary, the Ipswich Society for the Education of Females, and Scott, "The Ever-Widening Circle," 10–12, for examples.
89. For instance, Sara Evans writes, "Most important, however, the antislavery movement provided women with both an ideological and a practical training ground in political activism for democratic and egalitarian change." Sara M. Evans, *Born for Liberty: A History of Women in America* (New York: The Free Press, 1989), 80–81.

90. Stanton attended Troy Female Seminary, Lucy Stone and Antoinette Brown Blackwell attended Oberlin College, Abby Kelley attended the New England Friends Boarding School, and Susan B. Anthony attended a seminary in New York.

91. Woody, *A History of Women's Education*, I, 395; Barbara Miller Solomon, *In the Company of Educated Women: A History of Women and Higher Education in America* (New Haven, CT and London: Yale University Press, 1985), 23–24; Roger Geiger, "The Superior Education of Women," Unpublished section from "The Transformation of the Colleges," Presented at the 1995 meeting of the History of Education Society, 61.

92. Edward H. Clarke, *Sex in Education; Or, a Fair Chance for the Girls* (Boston: J. R. Osgood, 1873), quoted in Solomon, *In the Company of Educated Women*, 56.

93. Evans, *Born for Liberty*, 122.

94. Elizabeth Cady Stanton, "The Sixteenth Amendment," *Revolution* (April 29, 1869), 264–265. For discussion of this, see DuBois, *Feminism and Suffrage*.

95. Joe L. Dubbert, "Progressivism and the Masculinity Crisis," in *The American Man*, Elizabeth H. Pleck and Joseph H. Pleck, eds. (Englewood Cliffs, NJ: Prentice-Hall, 1980), 303–320; John Higham, ed., "The Reorientation of American Culture in the 1890s," *Writing American History: Essays on Modern Scholarship* (Bloomington: Indiana University Press, 1978), 73–102; Michael S. Kimmel, "The Contemporary 'Crisis' of Masculinity in Historical Perspective," in *The Making of Masculinities*, Harry Brod, ed. (Boston: Allen & Unwin, 1987), 121–154; James R. McGovern, "David Graham Phillips and the Virility Impulse of the Progressives," *New England Quarterly* 39 (September 1966): 334–355.

96. Blumin et al., *The Emergence of the Middle Class*, 290–295.

97. Peter G. Filene, *Him/Her/Self: Sex Roles in Modern America* (Baltimore: Johns Hopkins University Press, 1986), 73.

98. Alan Trachtenberg, *The Incorporation of America: Culture and Politics in the Gilded Age* (New York: Hill and Wang, 1982), 80.

99. Gail Bederman, *Manliness & Civilization: A Cultural History of Gender and Race in the United States, 1880–1917* (Chicago and London: The University of Chicago Press, 1995), 13.

100. Bederman, *Manliness & Civilization*, 14.

101. Mark C. Carnes, *Secret Ritual and Manhood in Victorian America* (New Haven, CT: Yale University Press, 1989); Jeffrey P. Hantover, "The Boy Scouts and the Validation of Masculinity," in *The American Man*, Elizabeth H. Pleck and Joseph H. Pleck, eds. (Englewood Cliffs, NJ: Prentice-Hall, 1980), 285–302.

102. Harvey Green, *Fit for America: Health, Fitness, Sport, and American Society* (Baltimore: Johns Hopkins University Press, 1986), 182–215; E. Anthony Rotundo, *American Manhood: Transformations in Masculinity from the Revolution to the Modern Era* (New York: Basic Books, 1993), 231–239, 251.

103. McCall, "Shall I Fetter Her Will?," 112.

104. U.S. Bureau of the Census statistics, quoted in Solomon, *In the Company of Educated Women*, 63.

105. Solomon, *In the Company of Educated Women*, 60.
106. Solomon, *In the Company of Educated Women*, ch. 4; Lynn D. Gordon, *Gender and Higher Education in the Progressive Era* (New Haven, CT and London: Yale University Press, 1990), ch. 1.
107. Solomon, *In the Company of Educated Women*, 60–61.

# Bibliography

## PRIMARY SOURCES

### Academy, Seminary, High School, and Other School
### Catalogs, Circulars, and Reports

*A Catalogue of the Abbot Female Seminary, Andover, Mass.* Andover, MA: Gould, Newman and Saxton, 1840.

Adams, William J. *Circular.* New York: n.p., 1829.

*Amherst Academy. Catalogue of the Trustees, Teachers and Students. For the Year Ending April, 1840.* Amherst?: n.p., 1840.

*Amherst Academy. Catalogue of the Trustees, Teachers and Students. For the Year Ending, April 1839.* Amherst?: n.p., 1839.

*Amherst Academy. Catalogue of the Trustees, Instructors and Students. For the Term Ending August 22, 1826.* Amherst: Carter and Adams, 1826.

Amherst Academy Circular, June 1832, Amherst College Archives.

Amherst Academy Broadsheet, 1827, Amherst College Archives.

*Amherst Academy.* Northampton: Shepard & Co., 1817.

*An Account of the High School for Girls, Boston: with A Catalogue of the Scholars. February, 1826.* Boston: Thomas B. Wait and Son, 1826.

*An Appeal to Parents for Female Education on Christian Principles; with a Prospectus of St. Mary's Hall, Green Bank, Burlington, New Jersey.* Burlington, NJ: J. L. Powell Missionary Press, 1837.

*Annual Catalogue of the Instructers* [sic] *and Pupils in the Newburgh (Late Mount Pleasant) Female Seminary. Situated in Newburgh, Orange Co. N.Y. for the Year preceding January 1st, 1837.* Newburgh, NY: J. D. Spalding, 1837.

*Annual Catalogue of the Officers, Teachers, and Pupils, of the Tuscaloosa Female Academy. For the Year 1832.* Tuscaloosa, AL: Expositor Office, 1833.

*Annual Circular, Etc., of the Geneva Female Seminary, under the Care of Mrs. Ricord.* Geneva: Merrill, printer, 1840.

*Annual Circular, Report and Catalogue of the Geneva Female Seminary, Under the Care of Mrs. Ricord. April, 1839.* Geneva: Ira Merrell, 1839.

*Arcade Ladies' Institute, Providence R.I.* Providence, RI: H. H. Brown, 1834?

*A Tabular View of the Order and Distribution of Studies Observed in the Respective Classes of the Hillsborough Female Seminary.* Hillsborough: n.p., 1826.

*Ballston Spa Female Seminary.* Albany, NY: Packard & Van Benthuysen, 1824.

*Brooklyn Collegiate Institute for Young Ladies; Brooklyn Heights, Opposite the City of New York.* New York: n.p., 1830.

*Candidates for Mount Holyoke Female Seminary.* South Hadley?: n.p., 1837?

*Catalogue and Circular of the LeRoy Female Seminary.* Rochester, NY: David Hoyt, 1840.

*Catalogue and Circular of the Ontario Female Seminary, Canandaigua; for the year ending June, 1839.* Canandaigua: C. Morse, 1839.

*Catalogue of Amherst Female Seminary, for the Year Ending August, 1835.* Amherst: J. S. and C. Adams, 1835.

*Catalogue of the Alabama Female Institute, Tuscaloosa, Ala. For the Year Ending July, 1838.* Tuscaloosa, AL: The Intelligencer Office, 1838.

*Catalogue of the Alabama Female Institute, Tuscaloosa, Ala. For the Year Ending July, 1837.* Tuscaloosa, AL: The Intelligencer Office, 1837.

*Catalogue of the Alabama Female Institute, Tuscaloosa, Ala. For the Year Ending 14th July, 1836.* Tuscaloosa, AL: Marmaduke J. Slade, 1836.

*Catalogue of the Greenfield High School for Young Ladies, for the Year 1836–37.* Greenfield, MA: Phelps & Ingersoll, 1837.

*Catalogue of the Greenfield High School for Young Ladies.* Greenfield, MA: Phelps & Ingersoll, 1830.

*Catalogue of the Instructors and Pupils in the New Haven Young Ladies' Institute, During its First Year.* New Haven, CT: n.p., 1830.

*Catalogue of the Instructors and Pupils of the Young Ladies' Institute, New Haven, Conn., for the Year Ending April, 1839.* New Haven, CT: Hitchcock & Stafford, 1839.

*Catalogue of the Instructors and Students in the Female Classical Seminary, Brookfield, Mass.* Brookfield, MA: E. & G. Merriam, 1826.

*Catalogue of the Members of the Female Seminary, under the care of Mr. Emerson, Commenced at Byfield and Continued at Saugus, Near Boston.* Boston: Crocker & Brewster, 1823.

*Catalogue of the Members of Uxbridge Female Seminary, During the Year Ending October 30, 1833.* Providence, RI: H. H. Brown, 1833.

*Catalogue of the Mount Vernon Female School. Containing the Names of the Trustees, Teachers and Pupils, in January, 1831.* Boston: T. R. Marvin, 1831.

*Catalogue of the Officers and Members of the Adams Female Academy, 1824, 1825, 1826.* Concord: Jacob B. moore, 1827.

*Catalogue of the Officers and Members of Bradford Academy, Bradford, Massachusetts, for the year ending November 20, 1840.* Haverhill, MA: Essex Banner Press, 1840.

*Catalogue of the Officers and Members of the Granville Female Academy, for the year ending February 22, 1838. Granville, Ohio.* Columbus, OH: Cutler and Pilsbury, 1838.

*Catalogue of the Officers and Members of Ipswich Female Seminary, for the Year Ending April, 1836.* Salem: The Landmark Office, 1836.

*Catalogue of the Officers and Members of Ipswich Female Seminary, for the Year Ending April, 1835.* Salem, MA: Palfray and Chapman, 1835.

*Catalogue of the Officers and Members of Ipswich Female Seminary, for the Year Ending October 1831.* Salem: Warwick Palfray, Jr., 1831.

*Catalogue of the Officers and Members of the Ipswich Female Academy.* Ipswich, MA: John Harris, Jr., 1829.

*Catalogue of the Officers and Members of the Sanderson Academy, 1828.* Ashfield?: n.p., 1828.

*Catalogue of the Officers and Members of the Seminary for Female Teachers, at Ipswich, Mass. For the Year Ending April 1839.* Salem: The Register Press, 1839.

*Catalogue of the Officers and Members of the Seminary for Female Teachers, at Ipswich, Mass. For the Year Ending April 1838.* Salem: William Ives & Co., 1838.

*Catalogue of the Officers and Members of the Seminary for Female Teachers, at Ipswich, Mass. for the Year Ending April 1837.* Salem, MA: Palfray and Chapman, 1837.

*Catalogue of the Officers and Members of the Utica Female Academy.* Utica, NY: Bennett, Backus, & Hawley, 1840.

*Catalogue of the Officers and Pupils of the Granville Female Seminary, for the Academical Year 1837.* Columbus, OH: Cutler and Pilsbury, 1837.

*Catalogue of the Officers and Students of Amenia Seminary, 1841–42.* Poughkeepsie, NY: Jackson & Schram, 1842.

*Catalogue of the Officers and Students of Bradford Academy, Bradford, Massachusetts, October, 1839.* Haverhill, MA: E. H. Safford, 1839.

*Catalogue of the Officers and Students of Bradford Academy, July, 1827.* Haverhill, MA: A. W. Thayer, 1827.

*Catalogue of the Officers and Students of the Townsend Female Seminary, for the Year Ending March, 1839.* Boston: John Putnam, 1839.

*Catalogue of the Officers and Students of the University of Alabama. For January, 1833.* Tuscaloosa, AL: E. Walker, 1833.

*Catalogue of the Officers and Students of the Wesleyan Academy, Wilbraham, Mass. Fall and Winter Terms, 1832–33.* Springfield, MA: G. & C. Merriam, 1833.

*Catalogue of the Officers, Teachers and Members of Day's Academy for Young Gentlemen, and Seminary for Young Ladies, from March 6, 1834, to January 14, 1835.* Boston: Beals and Greene, 1835.

*Catalogue of the South Hanover Female Seminary, for the Year Ending March, 1838.* South Hanover, NH: James Morrow, 1838.

*Catalogue of the Teachers and Pupils of the Buckland Female School, for the Term Ending March 2, 1830.* Greenfield, MA: Phelps & Ingersoll, 1830.

*Catalogue of the Teachers and Pupils of Byfield Seminary, 1821.* Newburyport: W. & J. Gilman, 1821.

*Catalogue of the Teachers and Pupils of the Female Seminary and Collateral School in Wethersfield, CT.* Wethersfield, CT: A. Francis, 1827.

*Catalogue of the Torringford Academy.* Torringford: n.p., 1831.

*Catalogue of the Trustees, Faculty and Students, of the Charleston College.* Charleston: A. E. Miller, 1824.

*Catalogue of the Trustees, Instructors, and Pupils in the Uxbridge Female Seminary, at Uxbridge, Mass.* Providence, RI: H. H. Brown, 1838.

*Catalogue of the Trustees, Instructers [sic], and Students of Amherst Academy, During the Quarter Ending August 9, 1825.* Greenfield?: n.p., 1825.

*Catalogue of the Trustees, Instructers [sic], and Students of Amherst Academy, During the Quarter Ending November 10, 1824.* Greenfield?: n.p., 1824.

*Catalogue of the Trustees, Instructers* [sic], *and Students of Amherst Academy, During the Quarter Ending Nov. 13, 1823*. Greenfield?: n.p., 1823.

*Catalogue of the Trustees, Instructers* [sic], *and Students of Amherst Academy*. Greenfield, MA: Denio & Phelps, 1822.

*Catalogue of the Trustees, Instructors, and Students of Amherst Academy*. Northampton: Shepard, 1821.

*Catalogue of the Trustees, Instructors, and Students of Amherst Academy*. Greenfield, MA: Denio & Phelps, 1818.

*Catalogue of the Trustees, Instructors, and Students of Amherst Academy*. Amherst?: W. W. Clapp, 1816.

*Catalogue of the Trustees, Instructors and Students of the Knoxville Female Academy*. Knoxville, TN: F. S. Heiskell, 1831.

*Catalogue of the Young Ladies' Seminary, in Keene, N.H. for the Year Ending October, 1832*. Keene, NH: J. & J. W. Prentiss, 1832.

*Circular of the Smithville Academy*. Providence, RI: n.p., 1838.

*Circular. To the Trustees and Patrons of the Female School connected with the Germantown Academy*. Germantown: n.p., 1832.

*Clinton Female Seminary*. Clinton: n.p., 1837.

Emerson, Joseph. *Prospectus of the Female Seminary, at Wethersfield, CT, Comprising a General Prospectus, Course of Instruction, Maxims of Education, and Regulations of the Seminary*. Wethersfield, CT: A. Francis, 1826.

*Fifth Annual Catalogue of the Teachers & Scholars in the Gothic Seminary, Northampton, Mass., September, 1840*. Northampton, MA: n.p., 1840.

*First and Second Catalogues of the Teachers and Members of Monticello Female Seminary, for the Years Ending 1839–40*. Alton: Parks' Book and Job Office, 1840.

*First Annual Catalogue of the Officers and Members of the Mount Holyoke Female Seminary, South Hadley, Mass*. South Hadley?: n.p., 1838.

*First Circular of the Oberlin Collegiate Institute, March 8, 1834*. Oberlin: n.p., 1834.

*Fourth Annual Catalogue of the Teachers and Scholars, in the Young Ladies' High School, Boston, July, 1831*. Boston: W. W. Clapp, 1831.

*General View of the Plan of Education Pursued at the Adams Female Academy*. Exeter, NH: Nathaniel S. Adams, printer, 1831.

*General View of the Principles and Design of the Mount Holyoke Female Seminary*. Boston: Perkins & Marvin, 1837.

*Greenfield High School for Young Ladies* [broadsheet]. Greenfield, MA: n.p., 1840.

*Laws for the Government of Newbern Academy, with the Plan of Education Annexed*. Newbern: S. Hall, 1823.

*Laws of the Raleigh Academy: With the Plan of Education Annexed*. Raleigh, NC: Gales & Seaton, 1811.

*Mount Holyoke Female Seminary. Preparation for Admission*. South Hadley: n.p., 1840.

*Mount Vernon Female Seminary*. Boston: n.p., 1836.

Neal, J. A. *An Essay on the Education and Genius of the Female Sex*. Philadelphia: Jacob Johnson, 1795.

*Newark Institute for Young Ladies*. Newark, NJ: n.p., 1826.

*Oberlin Collegiate Institute: 1st Circular, March 8, 1834.* Oberlin: n.p., 1834.

*Outline and Catalogue of the Steubenville Female Seminary for the Year Ending in October, 1840.* Steubenville, OH: n.p., 1840.

*Outline and Catalogue of the Steubenville Female Seminary for the Year Ending in September, 1839.* Steubenville, OH: n.p., 1839.

*Outline and Catalogue of the Steubenville Female Seminary for the Year Ending in September, 1838.* Steubenville, OH: n.p., 1838.

"Private Instruction" [Misses Watson; broadsheet]. Hartford, CT: n.p., 1834.

"Prospectus of St. Joseph's Academy for Young Ladies." Broadside: n.p., 1832.

*Prospectus of the Lexington Female Academy.* Lexington, KY: n.p., 1821.

*Prospectus of Mount Holyoke Female Seminary.* Boston: Perkins & Marvin, 1837.

*Prospectus of the Raleigh Academy, and Mrs. Hutchison's View of Female Education.* Raleigh, NC: C. White, 1835.

*Prospectus of the Regulations of the Alabama Female Athenaeum, Together with a List of its Superintendents, Officers and Teachers.* Tuscaloosa, AL: M. D. J. Slade, 1836.

*The Rise and Progress of the Young Ladies' Academy of Philadelphia: Containing an Account of a Number of Public Examinations & Commencements; The Charter and Bye-Laws; Likewise, A Number of Orations delivered by the Young Ladies, and Several by the Trustees of said Institution* [1789]. Philadelphia: Stewart & Cochran, 1794.

*Roxbury Female School.* Roxbury: n.p., 1830?

*Rules of the Female High-School in the City of New-York.* New York: Mercein & Penfold, 1826.

*Second Annual Catalogue of the Officers and Members of the Mount Holyoke Female Seminary, South Hadley, Mass., 1838–39.* South Hadley?: n.p., 1839.

*Semi-Annual Catalogue of the Trustees, Instructors and Students, of Westfield Academy, Fall and Winter Terms, 1840–41.* Northampton: J. Metcalf, 1841.

*Sparta Female Model School.* Augusta: B. Brantly, 1838.

*Statement of the Course of Study and Instruction Pursued at Washington College, Hartford, Connecticut; with a Catalogue of the Officers and Students.* Hartford, CT: P. Canfield, 1835.

*Terms and Conditions of the Boarding School for Female Education in Salem, N.C.* Salem: n.p., 1807.

*Terms and Conditions of the Institution for Female Education at Salem.* Salem: Blum & Son, 1840.

*Third Annual Catalogue of the Officers and Members of the Mount Holyoke Female Seminary. South Hadley, Mass., 1839–40.* South Hadley?: n.p., 1840.

"Uxbridge Female Seminary." Worcester, MA: n.p., 1837.

*The Granville Female Seminary, Chartered by the Legislature of Ohio: A Catalogue of its Officers and Pupils, for the Academical Year 1835. Together with General Information Respecting its Internal Arrangements.* Columbus, OH: The Temperance Advocate Office, 1836.

*Western Collegiate Institute, for Young Ladies.* Pittsburgh: n.p., 1837.

*Winchester Academical Institute for Young Ladies.* Winchester: n.p., 1835.

*Young Ladies' Boarding-School, Warrenton.* Broadside, May 19, 1809, Duke University Special Collections.

172    Bibliography

## Published Articles, Books, and Addresses

Abercrombie, James. *A Charge Delivered, after a Public Examination, On Friday, July 27, 1804, to the Senior Class of the Philadelphia Academy, upon their having Completed the Course of Study Prescribed by that Institution.* Philadelphia: H. Maxwell, 1804.

"Account of a Female School." *American Annals of Education and Instruction* II (April 1832): 209–217.

*A Course of Calisthenics for Young Ladies in Schools and Families With Some Remarks on Physical Education.* Hartford, CT: H. and F. J. Huntington, 1831.

Adams, Hannah. *A Memoir of Miss Hannah Adams, Written by Herself with Additional Notices by a Friend.* Boston: Gray and Bowen, 1832.

Addicks, Barbara O'Sullivan. *Essay on Education; in which the Subject is Treated as a Natural Science: In a Series of Short Familiar Lectures.* Philadelphia: Martin & Boden, 1831.

*Address Delivered 27th April, 1839, at the Opening of the Rutgers Female Institute, New-York.* New York: William Osborn, 1839.

"Address on Associations to Promote Education." *American Annals of Education and Instruction* V ( July 1835): 317.

"A Hint." *American Museum* 7 (1790): 208.

Alexander, William. *The History of Women.* Philadelphia: J. H. Dobelbower, 1796.

"An Account of a Work, entitled—A Defense of the Genius of Women. An Academical Discourse." *Universal Asylum* (November 1791): 323–324.

"An Address to the Ladies." *American Magazine* (March 1788): 241–245, 446.

Andrews, John. "Physical Education of Females." *American Annals of Education and Instruction* VI (February 1836): 85.

"An Error in Female Education." *American Annals of Education and Instruction* V (November 1835): 492–494.

"Annual Report Presented to the Trustees of the Elizabeth Female Academy, by Mrs. C. M. Thayer, Governess." *American Journal of Education* II (October 1827): 633–637.

"A Plan for a Matrimonial Lottery." *Lady's Magazine* ( June 1792): 173–175.

"Appropriate Exercise." *The Journal of Health* I (September 23, 1829): 22.

"Arithmetick Recommended to the Ladies." *Massachusetts Magazine* (February 1792): 106–107.

Badger, Catharine. *The Teacher's Last Lesson: A Memoir of Martha Whiting, Late of the Charlestown Female Seminary. Consisting Chiefly of Extracts from Her Journal, Interspersed with Reminiscences and Suggestive Reflections.* Boston: Gould and Lincoln, 1855.

Bailey, E. *Review of the Mayor's Report, on the Subject of Schools, So Far as it Relates to the High School for Girls.* Boston: Bowles & Dearborn, 1828.

Bancroft, Aaron. *Importance of Education, Illustrated in an Oration, Delivered Before the Trustees, Preceptors & Students of Leicester Academy, On the Fourth of July, 1806.* Worcester, MA: Thomas & Sturtevant, 1806.

Barnard, Henry, ed. *Memoirs of Teachers, Educators, and Promoters and Benefactors of Education, Literature, and Science. Reprinted from the American Journal of Education.* New York: F. C. Brownell, 1861.

————. "Educational Statistics of the United States in 1850." *American Journal of Education* I (1855): 368.

Beecher, Catharine. *The True Remedy for the Wrongs of Women.* Boston: Phillips, Sampson, 1851.

————. *The Evils Suffered by American Women and American Children: The Causes and the Remedy.* New York: Harper & Bros., 1846.

————. *Essay on Slavery and Abolitionism with Reference to the Duty of American Females.* Philadelphia: Henry Perkins, 1837.

————. *An Essay on the Education of Female Teachers.* New York: Van Nostrand & Dwight, 1835.

————. *Suggestions Respecting Improvements in Education. Presented to the Trustees of the Hartford Female Seminary, and published at their Request.* Hartford, CT: Packard & Butler, 1829.

————. "Female Education." *American Journal of Education* II (April 1827): 219–223; (May 1827): 264–269, 484, 739.

Bell, John. "Physical Education of Girls." *The Journal of Health* I (September 9, 1829): 15.

"Benefits of Exercise." *The Journal of Health* II (November 30, 1830): 91–92.

"Boston High School for Girls." *American Journal of Education* II (March 1827): 184–187; (April 1827): 206–210.

"Boston High School for Girls." *American Journal of Education* I ( July 1826): 380–381; (February 1826): 96–105.

"Boston Monitorial School." *American Journal of Education* I ( January 1826): 29.

Budden, Maria. *Thoughts on Domestic Education; the Result of Experience. By a Mother.* London: Charles Knight, 1826.

"Buffalo High School." *American Journals of Education and Instruction* III (April 1828): 233–235.

Burnside, Samuel M. "Speech on the Opening of Two New Female Schools." Worcester, MA: n.p., 1833.

Burroughs, Charles. *An Address on Female Education, Delivered in Portsmouth, New-Hampshire, October 26, 1827.* Portsmouth: Childs and March, 1827.

Butler, Charles. *The American Lady.* Philadelphia: Hogan & Thompson, 1836.

"Calisthenics." *The Journal of Health* II (February 23, 1831): 190.

Carey, Mathew. *Education.* Philadelphia: n.p., 1828.

Chandler, Daniel. *An Address on Female Education, Delivered before the Demosthenian & Phi Kappa Societies, on the Day after Commencement, in the University of Georgia, by Daniel Chandler, Esq.* Washington, GA: William A. Mercer, 1835.

"Characteristic Differences of the Male and Female of the Human Species." *New York Magazine* (June 1790): 336–338.

Chevalier, Michael. *Society, Manners and Politics in the United States: Being a Series of Letters on North America.* Boston: Weeks, Jordan & Company, 1839.

Clarke, Edward. *The Building of a Brain.* Boston: J. R. Osgood, 1874.

————. *Sex in Education, or A Fair Chance for Girls.* Boston: J. R. Osgood, 1873.

"Columbia Female Institute." *American Annals of Education and Instruction* VIII (October 1838): 435–441.

"Constitution, Female Literary Association." *The Liberator* (December 3, 1831).

Cotting, John Ruggles. *An Address Delivered at the Female Classical Seminary, Brookfield, Jan. 22, 1827, at the Close of the Introductory Lecture at the Winter Course on Natural Science.* Brookfield, MA: E. G. Merriam, 1827.

"Course of Education in the New-York High-School." *American Journal of Education* I (January 1826): 23–29.

"Curious Dissertation on the Valuable Advantages of a Liberal Education." *New Jersey Magazine* (January 1787): 49–55.

Darwin, Erasmus. *A Plan for the Conduct of Female Education in Boarding Schools.* London: J. Drewry, 1797.

Day, Jeremiah and James Kingsley. "Reports on the Course of Instruction in Yale College; by a Committee of the Corporation and the Academical Faculty." In *American Journal of Science and Arts* 15 (January 1829): 297–351.

DeRosenstein, M. "Vindication of the Study of Polite literature." *New York Magazine* (May 1793): 282–288.

"Destination of Woman." *American Annals of Education and Instruction* VI (November 1836): 503–505.

"Dialogue on Female Education." *Portico* (1816): 210–215.

"Disasters Overcome." *Lowell Offering* 2 (1841): 289–297.

Doggett, Simeon. "A Discourse on Education, Delivered at the Dedication and Opening of Bristol Academy, the 18th Day of July, A.D. 1796" [1796]. In *Essays on Education in the Early Republic*, Frederick Rudolph, ed. Cambridge: Belknap Press of Harvard University Press, 1965, 147–165.

"Domestic Seminary for Young Ladies." *American Annals of Education and Instruction* IV (November 1834): 498–503.

Doolittle, Julia. *Extent of Individual Influence: First Prize Essay of the Albany Female Academy, July, 1835.* Albany, NY: Packard & Van Benthuysen, 1835.

Douglass, Sarah Mapps. "Address." *The Liberator* (July 21, 1832).

"Early Education of Females." *American Journal of Education* III (April 1828): 209–214.

Edgeworth, Maria. *Letters for Literary Ladies. To which is added, an essay on the noble science of self-justification.* George Town: Joseph Milligan, 1810.

Edgeworth, Maria and Richard Lovell Edgeworth. *Practical Education.* New York: George F. Hopkins, 1801.

"Education." *American Magazine* (December 1787): 25–27.

"Education." *The Port Folio* (May 1803): 174.

"Education." *The Port Folio* (April 1802): 116.

"Education in the State of New-York: Extract from Gov. Clinton's Message, Jan. 3, 1826." *American Journal of Education* I (January 1826): 58–59.

"Education of Females." *American Journal of Education* II (June 1827): 339–343; (July 1827): 423–428; (August 1827): 481–487; (November 1827): 549–555; (December 1827): 604–609.

"Education of Females." *American Journal of Education* I (July 1826): 350.

"Education of Females: Domestic Management." *American Journal of Education* III (May 1828): 343–346.

"Education of Females: Intellectual Instruction." *American Journal of Education* II (November 1827), 676–682; (December 1827): 734–742.

"Education of Females: Motives to Application." *American Journal of Education* II (September 1827), 549–550.

"Effects of Emulation." *American Annals of Education and Instruction* IV (August 1834): 349–353.

Emerson, Joseph. *Letter to a Class of Young Ladies, Upon the Study of the History of the United States.* Boston: Crocker & Brewster, 1828.

———. *Mr. Emerson's Recitation Lectures, Upon the Acquisition and Communication of Thought; Proposed for the Ensuing Season.* Wethersfield, CT: A. Francis, 1826.

———. *Female Education. A Discourse, Delivered at the Dedication of the Seminary Hall in Saugus, Jan. 14, 1822.* Boston: Samuel T. Armstrong, and Crocker & Brewster, 1823.

Emerson, Ralph. *Life of Rev. Joseph Emerson, Pastor of the Third Congregational Church in Beverly, Ma. and subsequently Principal of a Female Seminary.* Boston: Crocker & Brewster, 1834.

"Emulation." *American Annals of Education and Instruction* VI (March 1836): 108–110.

"Errors in Physical Education." *American Annals of Education and Instruction* VI (March 1836): 105–110.

"Essay on Education." *New York Magazine* (January 1790): 40–41.

"Evils in Female Education." *American Annals of Education and Instruction* I (December 1831): 560–564.

"Extremes in Female Education." *American Annals of Education and Instruction* V (November 1835), 464–467.

Felt, Joseph B. *To the Friends and Patrons of Ipswich Female Seminary.* Ipswich, MA: n.p., 1834.

"Female Academy at Sturgeonville, Virginia." *American Journal of Education* III (November 1828): 618–620.

"Female Accomplishments: A Dialogue." *The Parlour Companion* (August 1817): 123, 131.

"Female Education." *American Annals of Education and Instruction* VIII (April 1838): 172–175.

"Female Education." *American Annals of Education and Instruction* VII (May 1837): 219–222; (October 1837): 447–451.

"Female Education." *American Annals of Education and Instruction* VI (March 1836): 100–105.

"Female Education." *American Annals of Education and Instruction* V (June 1835): 262–266; (July 1835): 314–316; (August 1835): 360–363; (September 1835): 415–417.

"Female Education." *American Annals of Education and Instruction* IV (July 1834): 299–301.

"Female Education." *American Journal of Education* III (September 1828): 517–527; (November 1828): 648–652.

"Female Education." *The American Quarterly Observer* III (October 1834): 384–385.

"Female Education in the Last Century." *American Annals of Education and Instruction* I (November 1831): 522–526.

"Female Education Should Be Thorough." *American Annals of Education and Instruction* VIII (September 1838): 390–392.

"Female Education Subsequent to the Period of Going to School." *American Journal of Education* III (July 1828): 416–420.

"Female High-School of Boston." *American Journal of Education* I ( January 1826): 61.

"Female High-School of New-York." *American Journal of Education* I ( January 1826): 59–60.

"Female Influence." *New York Magazine* (May 1795): 297–304.

"Female Learning." *Weekly Magazine of Original Essays, Fugitive Piecs, and Interesting Intelligence* (May 1798): 60.

"Female Patriotism and Fortitude." *Massachusetts Magazine* (March 1791): 165–166.

"Female Seminaries." *American Annals of Education and Instruction* VI (May 1836): 235–236.

"Female Vanity." *New York Magazine* (December 1790): 694–695.

Ferris, Isaac. *Address Delivered 27th April, 1839, At the Opening of the Rutgers Female Institute, New-York.* New York: William Osborn, 1839.

*Fifth Annual Report of the Young Ladies' Association of the New-Hampton Female Seminary, for the Promotion of Literature and Missions; with the Constitution, etc. 1838–39.* Boston: John Putnam, 1839.

*Fourth Annual Report of the Young Ladies' Association of the New-Hampton Female Seminary, for the Promotion of Literature and Missions; with the Constitution, etc. 1837–38.* Boston: Freeman and Bolles, 1838.

Franklin, Benjamin. "Proposals Relating to the Education of Youth in Pensilvania" [1749]. In *The Age of the Academies.* Theodore R. Sizer, ed. New York: Teachers College, 1964, 68–76.

"Fundamental Principles of Female Education." *American Annals of Education* (February 1834): 85–87.

Gallaudet, Thomas H. *An Address on Female Education, delivered, November 21st, 1827, at the Opening of the Edifice erected for the accommodation of the Hartford Female Seminary.* Hartford, CT: H. & F.J. Huntington, 1828.

Garnett, James M. *A Discourse on scholastic Reforms, and Amendments in the Modes and General Scope of Parental Instruction. Delivered before the Fredericksburg Lyceum, September 28, 1832.* Richmond, VA: Thomas W. White, 1833.

———. "An Address on the subject of Literary Associations to promote Education; delivered before the Institute of Hampden Sidney College, Va., at their last Commencement." *American Annals of Education and Instruction* V ( July 1835): 317–320.

———. *Lectures on Female Education, Comprising the First and Second Series of a Course Delivered to Mrs. Garnett's Pupils, At Elm-Wood, Essex County, Virginia. By James M. Garnett. To Which is Annexed, The Gossip's Manual.* 3rd ed. Richmond, VA: Thomas W. White, 1825.

"Georgia Female College." *American Annals of Education and Instruction* VIII (November 1838): 523–524.

Gilman, Caroline, ed. *Letters of Eliza Wilkinson, during the Invasion and Possession of Charleston, S.C., by the British in the Revolutionary War* [1839]. New York Times & New York: Arno Press, 1969.

Gisborne, Thomas. *An Enquiry into the Duties of the Female Sex.* London: T. Cadell and W. Davies, 1798.

Grant, Anne Mac Vicar. *Sketches of Intellectual Education, and Hints on Domestic Economy, Addressed to Mothers: with an Appendix, containing An Essay on the Instruction of the Poor.* Baltimore: Edward J. Coale, 1813.

Grant, Zilpah. "Benefits of Female Education." n.p., 1836.

Gregory, George. "Miscellaneous Observations on the History of the Female Sex." *Essays Historical and Moral.* London: J. Johnson, 1785.

"Gymnastic Exercises for Females." *American Journal of Education* I (November 1826): 698–701.

Hall, Baynard R. *An Address Delivered to the Young Ladies of the Spring-Villa Seminary, at Bordentown, N.J. At the Distribution of the Annual Medal and Premiums: On the Evening of the 29th of August, 1839.* Burlington, NJ: Powell & George, 1839.

Hamilton, Elizabeth. *Letters on Education.* Dublin: H. Colbert, 1801.

"Hartford Female Seminary. *American Annals of Education and Instruction* II (January 1832): 62–68; (April 1832): 218–225.

"Hartford Female Seminary." *American Journal of Education* III (August 1828): 461–465.

Hays, Mary. "On the Independence and Dignity of the Female Sex." *New York Magazine* (April 1796): 202–203.

"High School for Young Ladies at Greenfield, Mass." *American Journal of Education* III (August 1828): 488–489.

"High School of Buffalo, New-York." *American Journal of Education* III (April 1828): 233–235.

"Hints on Reading." *Lady's Magazine* (March 1793): 171–173.

Hobson, John. *Prospectus of a Plan of Instruction for the Young of Both Sexes, Including a Course of Liberal Education for Each.* Philadelphia: D. Hogan, printer, 1799.

Hopkins, Alice L. *A Reminiscence of Miss A. D. Doolittle and the Rochester Female Academy.* Rochester, NY: n.p., n.d.

"Improvements Suggested in Female Education." *Monthly Magazine* (March 1797), reprint in *New York Magazine* (August 1797): 405–408.

"Inconveniences from a Too Loving Wife." *New York Magazine* (December 1794): 754–757.

"Inconvenience of a Learned Wife." *The Port Folio* (September 1802): 275–276.

"Insanity from Excessive Study." *American Annals of Education and Instruction* III (September 1833): 425.

"Instructions Preparatory to the Married State." *New York Magazine* (July 1797): 374.

Irving, John T. *Address Delivered on the Opening of the New-York High-School for Females, January 31, 1826.* New York: William A. Mercein, 1826.

Kames, Henry Home, Lord. *Six Sketches on the History of Man* [1774]. Philadelphia, n.p., 1776.

Knox, Samuel. "An Essay on the Best System of Liberal Education, Adapted to the Genius of the Government of the United States. Comprehending also, an

Uniform General Plan for Instituting and Conducting Public Schools, in This Country, on Principles of the Most Extensive Utility. To Which Is Prefixed, an Address to the Legislature of Maryland on That Subject" [1799]. In *Essays on Education in the Early Republic*, Frederick Rudolph, ed. Cambridge: Belknap Press of Harvard University Press, 1965, 271–372.

"Ladies' Association for Educating Females, in Illinois." *American Annals of Education and Instruction* VII (April 1837): 185–186.

"Ladies' Association for the Education of Female Teachers." *American Annals of Education and Instruction* V (January 1835): 81–84.

La Maissoneuve, Antoinette Legroing. "Female Education." *New York Mirror and Ladies' Literary Gazette* (August 1823): 4, 10, 18–19, 27, 36.

Lansing, Marion, ed. *Mary Lyon Through Her Letters*. Boston: Books, Inc., 1937.

Lee, Hannah Farnham Sawyer. *The Contrast: or Modes of Education*. Boston: Whipples and Damrell, 1837.

"Letter from a Young Lady." *United States Magazine* (June 1779): 261–264.

Locke, John. "Thoughts on Education." In *The Educational Writings*, James Axtell, ed. Cambridge: Cambridge University Press, 1968.

Ludlow, John. *An Address Delivered at the Opening of the New Female Academy in Albany, May 12, 1834, By John Ludlow, D.D., President of the Board of Trustees, with an Appendix*. Albany, NY: Packard & Van Benthuysen, 1834.

Lyon, Mary. *Female Education. Tendencies of the Principles Embraced, and the System Adopted in the Mount Holyoke Female Seminary*. South Hadley?: n.p., 1839.

Magaw, Samuel. *An Address Delivered in the Young Ladies Academy, at Philadelphia, on February 8th, 1787. At the Close of a Public Examination*. Philadelphia: Thomas Dobson, 1787.

Marks, Elias. *Hints on Female Education, with an Outline of an Institution for the Education of Females, Termed the So. Ca. Female Institute; under the direction of Dr. Elias Marks*. Columbia, SC: David W. Sims, 1828.

Mayes, Daniel. *An Address Delivered on the First Anniversary of Van Doren's Collegiate Institute for Young Ladies, in the City of Lexington, Ky. On the Last Thursday of July, 1832*. Lexington, KY: Finnell & Herndon, 1832.

M. F. B. "Answer to a Father's Inquiries relative to the Education of Daughters." *The New England Quarterly Magazine* (December 1802): 156.

Millar, John. "The Origins of the Distinction of Ranks; or, An Inquiry into the Circumstances which give Rise to Influence and Authority in the Different Members of Society" [1771]. In *John Millar of Glasgow, 1735–1801*, William C. Lehmann, ed. London and New York: Cambridge University Press, 1960.

"Miss Beecher's Essay on the Education of Female Teachers." *American Annals of Education and Instruction* V (July 1835): 275–278.

"Miss Sedgewick's *Tales*." *The North American Review* (October 1837): 480–481.

More, Hannah. "Essays" [1792]. Reprinted in *The Ladies' Companion*. Worcester, MA: The Spy Office, 1824.

More, Hannah. "Thoughts on the Cultivation of the Heart and Temper in the Education of Daughters" [1794]. In *The Lady's Companion*. Worcester, MA: The Spy Office, 1824, 92–98.

Morse, Jedediah. *Geography Made Easy* [1784]. Boston: J. T. Buckingham, 1806.

"Motives to Study in the Ipswich Female Seminary." *American Annals of Education and Instruction* III (February 1833): 75–80.

Mott, Abigail. *Observations on the Importance of Female Education, and Maternal Instruction, with their Beneficial Influence on Society. By a Mother.* New York: Mahlon Day, 1825.

"Mount Holyoke Female Seminary." *American Annals of Education and Instruction* V (August 1835): 375–376.

"Mount Holyoke Female Seminary." *Boston Recorder* (May 13, 1836).

Murray, Judith Sargent. In *The Gleaner* [1798], Nina Baym, ed. Schenectady, NY: Union College Press, 1992.

Neal, James A. *An Essay on the Education and Genius of the Female Sex.* Philadelphia: Jacob Johnson 1795.

"Neglect of Females." *American Annals of Education and Instruction* VI (November 1836): 523.

"New York High School." *American Journal of Education* II (January 1827): 58–60.

*Notes on the Sayings and Doings of Dr. Lacey and His Three Friends; As Partly Set Forth in their Celebrated Report, with its Numerous Affidavits and Other Appendages.* Albany, NY: n.p., 1830.

"Oberlin Collegiate Institute." *American Annals of Education and Instruction* VIII (October 1838): 477.

"Observer, No. 1." *The Key* (January 27, 1798): 17.

"Of the Unhappiness of Women." *New York Magazine* (October 1797): 525–528.

Ogden, John Cossens. *The Female Guide.* Concord, NH: George Hough, 1793.

"On Compulsory Laws respecting Marriage." *New York Magazine* (December 1791): 715–717.

"On Family Ambition." *Lady's Magazine* (August 1792): 119–123.

"On Female Authorship." *Lady's Magazine* (January 1793): 68–72.

"On Female Education." *American Journal of Education* III (May 1828): 162.

"On Female Education." *New York Magazine* (September 1794): 569–570.

"On Marriage." *New York Magazine* (December 1796): 656–657.

"On Matrimonial Obedience." *Lady's Magazine* (June 1792): 64–67.

"Ontario Female Seminary." *American Journal of Education* III (July 1828): 426–427.

"On the Choice of a Wife." *Literary Museum* (June 1797): 295–299.

"On the Domestic Character of Women." *New York Magazine* (November 1796): 600–602.

"On the Education and Studies of Women." *New York Magazine* (October 1797): 540–543.

"On the Independence and Dignity of the Female Sex." *New York Magazine* (April 1796): 202–203.

"On the Policy of Elevating the Standard of Female Education, Addressed to the American Lyceum, May, 1834." *American Annals of Education and Instruction* IV (August 1834): 361–364.

"On the Study of the Arts and Sciences." *New York Magazine* (June 1795): 363–364.

"On the Supposed Superiority of the Masculine Understanding." *Universal Asylum* (July 1791): 9–11.

"On Women." *Massachusetts Magazine* (June 1791): 333.

"On Women." *New York Magazine* (January 1790): 16–17.

"Outline of a Plan of Instruction for the Young of Both Sexes, particularly Females, submitted to the Reflection of the Intelligent and the Candid." *Weekly Magazine of Original Essays, Fugitive Pieces, and Interesting Intelligence* (August 4, 1798): 12–15; (August 11, 1798): 37–41.

Phelps, Almira Hart Lincoln. *The Female Student; Or, Lectures to Young Ladies on Female Education,* 2nd ed. New York: Leavitt, Lord & Co., 1836.

Phelps, Almira Hart Lincoln. *Caroline Westerley; or, The Young Traveller from Ohio. Containing the Letters of a Young Lady of Seventeen. Written to Her Sister.* New York: J. & J. Harper, 1833.

"Physical Education of Females." *American Annals of Education and Instruction* VI (February 1836): 84–86.

"Popular Education." *North American Review* 36 (January 1833): 82–83.

"Prefatory Address." *American Journal of Education* I (January 1827): 1–13.

"Progress of Female Education." *American Annals of Education* I (September 1830): 97–99.

"Proof of the Existence of a Reasonable Woman." *New York Magazine* (February 1797): 93–95.

"Prospectus." *American Journal of Education* I (January 1826), 1–4.

"Providence High School." *American Journal of Education* III (July 1828): 427–430.

Randall, Anne Frances. *A Letter to the Women of England, on the Injustice of Mental Subordination. With Anecdotes.* London: T. N. Longman & O. Rees, 1799.

"Reflections on Women, and on the Advantages which they would derive from the Cultivation of Letters." *New York Magazine* (February 1790): 89–90.

"Retrospect." *American Journal of Education* II (December 1827): 754–760.

"Review." *American Journal of Education* II (July 1827): 428–481.

"Review of Charles Londe, *Medical Gymnastics.*" *American Journal of Education* I (April 1826): 235–239.

*Reviewer of Mrs. Emma Willard Reviewed.* Philadelphia: C. Sherman & Co., 1839.

"Rights of Women." *New York Magazine* (December 1791): 713–714.

"Rochester Female Seminary." *Rochester Daily Democrat* (March 31, 1834): 2.

Rudolph, Frederick, ed. *Essays on Education in the Early Republic.* Cambridge: Belknap Press of Harvard University Press, 1965.

Rush, Benjamin. "Thoughts Upon Female Education, Accommodated to the Present State of Society, Manners, and Government in the United States of America" [1787]. In *Essays on Education in the Early Republic,* Frederick Rudolph, ed. Cambridge: Belknap Press of Harvard University Press, 1965, 25–40.

———. "A Plan for the Establishment of Public Schools and the Diffusion of Knowledge in Pennsylvania; To Which Are Added, Thoughts upon the Mode of Education, Proper in a Republic. Addressed to the Legislature and Citizens of the State" [1786]. In *Essays on Education in the Early Republic,* Frederick Rudolph, ed. Cambridge: Belknap Press of Harvard University Press, 1965, 1–23.

"Scheme for Increasing the Power of the Fair Sex." *Lady's Magazine* (June 1792): 22–24.

Scott, John W. *An Address on Female Education, Delivered At the Close of the Summer Session for 1840, of the Steubenville Female Seminary, in Presence of its Pupils and Patrons.* Steubenville, OH: n.p., 1840.

*Second Annual Report of the Young Ladies' Association of the New-Hampton Female Seminary, for the Promotion of Literature and Missions; with the Constitution, etc. 1834–5.* Boston: Freeman and Bolles, 1836.

Sedgwick, Catharine M. *Means and Ends, or Self-Training.* Boston: March, Capen, Lyon & Webb, 1839.

"Self Improvement an Important Part of Female Education." *American Journal of Education* III (March 1828): 161–166.

"Seminary for Female Teachers, at Ipswich, Mass." *American Annals of Education and Instruction* III (February 1833): 69–75.

"Seminary for Female Teachers." *American Annals of Education and Instruction* I (August 1831): 341–346.

S. F. W. "Female Education: Extract from a Letter to the Editor of *Ladies' Magazine.*" *American Annals of Education and Instruction* IV (July 1834): 300.

*Sixth Annual Report of the Young Ladies' Association of the New-Hampton Female Seminary, for the Promotion of Literature and Missions; with the Constitution, etc. 1839–40.* Boston: Putnam & Hewes, 1840.

Smith, Samuel. "Remarks on Education: Illustrating the Close Connection Between Virtue and Wisdom. To Which Is Annexed a System of Liberal Education. Which, Having Received the Premium Awarded by the American Philosophical Society, December 15th, 1797, Is Now Published by Their Order" [1797]. In *Essays on Education in the Early Republic,* Frederick Rudolph, ed. Cambridge: Belknap Press of Harvard University Press, 1965, 167–223.

"Society for the Education of Females." Ipswich Female Seminary: n.p., 1835.

Stanton, Elizabeth Cady. "The Sixteenth Amendment." *Revolution* (April 29, 1869): 264–265.

Stith, Townshend (Mrs.). *Thoughts on Female Education.* Philadelphia: Clark & Raser, 1831.

Stocking, S. S. *An Address, Delivered Before the Young Ladies' Literary Society of the Wesleyan Academy, June 8, 1836.* Boston: David H. Ela, 1836.

Story, Joseph. *A Discourse pronounced before the Phi Beta Kappa Society, at the Anniversary Celebration, on the Thirty-first Day of August, 1826.* Boston: Hilliard, Gray, Little, and Wilkins, 1826.

"Suggestions to Parents on Female Education: Accomplishments." *American Journal of Education* III (May 1828): 276–283.

"Suggestions to Parents: Physical Education." *American Journal of Education* I (April 1826): 235–239.

Swanwick, John. *Thoughts on Education, Addressed to the Visitors of the Young Ladies' Academy in Philadelphia, October 31, 1787.* Philadelphia: Thomas Dobson, 1787.

"The Convenience of a Scolding Wife." *Lady's Magazine* (December 1792): 18–19.

"The History of Frivola: An important Lesson for the Fair Sex." *New York Magazine* (March 1790): 160–161.

"The Influence of the Female Sex on the Enjoyments of Social Life." *Universal Asylum* (March 1790): 153–154.

*The Lady's Companion, Containing, First, Politeness of Manners and Behavior, from the French of the Abbe de Bellegarde. Second, Fenelon on Education—Third, Miss More's Essays—Fourth, Dean Swift's Letter to a Young Lady Newly Married—Fifth, Moore's Fables for the Female Sex. Carefully selected and revised by a Lady.* Worcester, MA: The Spy Office, 1824.

*The Lady's Pocket Library.* Philadelphia: Mathew Carey, 1792.

"The Plague of a Learned Wife." *Weekly Magazine of Original Essays, Fugitive Pieces, and Interesting Intelligence* (May 1798): 89–90.

"The Propriety of Meliorating the Condition of Women in Civilized Societies, Considered." *American Museum* 6 (1789): 248–249.

"The System of Public Education, Adopted by the Town of Boston, October 15, 1789." *New York Magazine* (January 1790): 52.

"The Widow's Son." *Lowell Offering* 2 (1841): 246–250.

*Third Annual Report of the Young Ladies' Association of the New-Hampton Female Seminary, for the Promotion of Literature and Missions; with the Constitution, etc. 1835–36.* Boston: John Putnam, 1837.

"Thoughts on Old Maids." *Lady's Magazine* (July 1792): 60–62.

"Thoughts on the Education of Females." *American Journal of Education* I (June 1826): 349–352; (July 1826): 401–402.

"Thoughts on Women." *Lady's Magazine* (1792): 111–113.

"'Tis Education Forms the Female Mind." *Massachusetts Magazine* (February 1793): 92–93.

Trollope, Anthony. *North America* [1861]. New York: St. Martin's Press, 1986.

"Variety in Exercise." *The Journal of Health* I (April 28, 1830): 243–244.

Webster, Noah. "On the Education of Youth in America" [1790]. In *Essays on Education in the Early Republic*, Frederick Rudolph, ed. Cambridge: Belknap Press of Harvard University Press, 1965, 41–77.

"Western Female Institute." *American Annals of Education and Instruction* III (August 1833): 380–381.

Willard, Emma. *An Address to the Public: Particularly to the Members of the Legislature of New York, Proposing a Plan for Improving Female Education* [1819]. Middlebury, VT: Middlebury College, 1918.

Worcester, Leonard. *An Address on Female Education. Delivered at Newark (N.J.) March 28, 1832.* Newark?: n.p., 1832.

Worcester, Samuel. *The Christian Mourning with Hope. A Sermon, Delivered at Beverly, Nov. 14, 1808, on Occasion of the Death of Mrs. Eleanor Emerson, Late Consort of the Rev. Joseph Emerson.* Boston: Lincoln & Edmands, 1809.

## Manuscript Collections

Adams Female Academy Records, Mount Holyoke College Archives and Special Collections.

Albany Female Seminary Trustees' Minute Book, 1827–1849, New York State Library.

Amherst Academy Correspondence, Amherst College Archives.

Atkinson Academy Record Book, American Antiquarian Society.
Brownrigg Family Papers, Southern History Collection, UNC.
Buckland Female School Papers, Mount Holyoke College Archives and Special Collections.
Cameron Family Papers, Southern History Collection, UNC.
Carolina Eliza Clitherall Diaries, Southern History Collection, UNC.
Joseph Emerson Papers, Mount Holyoke College Archives and Special Collections.
Galloway Family Papers, Southern History Collection, UNC.
Zilpah Grant Papers, Mount Holyoke College Archives and Special Collections.
Peter Hagner Papers, Southern History Collection, UNC.
Pardon Hathaway Papers, Cornell University Archives.
John de Berniere Hooper Papers, Southern Historical Collection, UNC.
Chiliab Howe Papers, Southern Historical Collection, UNC.
Juliana Howell Journal, Cornell University Archives.
Susan Nye Hutchison Diary, Southern Historical Collection, UNC.
John Keep Papers, Oberlin College Archives.
Ipswich Female Seminary Papers and Correspondence, Mount Holyoke College Archives.
Jones Family Papers, Southern Historical Collection, UNC.
Keenan Family Papers, Southern Historical Collection, UNC.
Lambdin Family Papers, American Antiquarian Society.
Lenoir Family Papers, Southern Historical Collection, UNC.
Lincolnton Female Academy Records, Duke University Special Collections.
Mary Lyon Papers, Mount Holyoke College Archives and Special Collections.
McMullen Family Papers, Duke University Special Collections.
Jacob Mordecai Papers, Duke University Special Collections.
Mount Holyoke Correspondence and Records, Mount Holyoke College Archives.
Parish Family Papers, Southern Historical Collection, UNC.
Nathaniel Tracy Sheafe Papers, Cornell University Archives.
Skinner Family Papers, Southern Historical Collection, UNC.
Steele Family Papers, Southern Historical Collection, UNC.
Stephens Family Papers, Southern Historical Collection, UNC.
Treasurer's Office Correspondence, Oberlin College Archives.
Sukey Vickery Diary, American Antiquarian Society.
Peter Wainwright, Jr. Papers, Duke Special Collections.
Henry Watson, Jr. Papers, Duke Special Collections.
Emma Willard Papers, Sophia Smith Collection.

## Newspapers, Magazines, and Journals

*American Annals of Education and Instruction*
*American Journal of Education*
*American Magazine*
*American Magazine of Wonders*
*American Museum*

*Boston Weekly Magazine*
*Evening Fireside*
*Female Advocate*
*Gentlemen and Ladies Town and Country Magazine*
*Godey's Lady's Book*
*The Key*
*The Knickerbocker*
*Lady's Magazine*
*Massachusetts Magazine*
*Massachusetts Mercury and New-England Palladium*
*New England Quarterly Magazine*
*New Jersey Magazine*
*New York Magazine*
*New York Weekly Advocate*
*North Carolina Minerva and Fayetteville Advertiser*
*The Pennsylvania Gazette*
*Philadelphia's Gazette*
*Raleigh Register*
*Rochester Daily Advertiser and Telegraph*
*Rochester Daily Democrat*
*Royal American Magazine*
*State Gazette of North Carolina*
*State Gazette of South Carolina*
*Universal Asylum*
*Weekly Magazine of Original Essays, Fugitive Pieces, and Interesting Intelligence*


SECONDARY SOURCES

Allmendinger Jr., David F. "Mount Holyoke Students Encounter the Need for Life Planning, 1837–1850." *History of Education Quarterly* 19 (Spring 1979): 27–46.

Amireh, Amal. *The Factory Girl and the Seamstress: Imagining Gender and Class in Nineteenth Century American Fiction.* New York and London: Garland Publishing, 2000.

Antler, Joyce. " 'After College, What?': New Graduates and the Family Claim." *American Quarterly* 32 (Fall 1980): 409–434.

Apple, Michael W. "Teaching and 'Women's Work': A Comparative Historical and Ideological Analysis." *Teachers College Record* 86 (Spring 1985): 457–473.

Baker, Paula. "The Domestication of Politics: Women and American Political Society, 1780–1920." *American Historical Review* 89 (June 1984): 620–647.

Baym, Nina. "Women and the Republic: Emma Willard's Rhetoric of History." *American Quarterly* 43 (March 1991): 1–23.

Beadie, Nancy. " 'To Improve Every Leisure Moment': The Significance of Academy Attendance in the Mid-Nineteenth Century." Paper Presented at the History of Education Society Annual Meeting (October 1999).

————. "Female Students and Denominational Affiliation: Sources of Success and Variation among Nineteenth-Century Academies." *American Journal of Education* 107 (February 1999): 75–115.

————. "Emma Willard's Idea Put to the Test: The Consequences of State Support of Female Education in New York, 1819–67." *History of Education Quarterly* 33 (Winter 1993): 543–562.

Bederman, Gail. *Manliness & Civilization: A Cultural History of Gender and Race in the United States, 1880–1917.* Chicago and London: The University of Chicago Press, 1995.

Bernard, Richard M. and Maris A. Vinovskis. "The Female School Teacher in Ante-Bellum Massachusetts." *Journal of Social History* 10 (March 1977): 332–345.

Biklen, Sari Knopp. *School Work: Gender and the Cultural Construction of Teaching.* New York: Teachers College Press, 1995.

Bledstein, Burton S. and Robert D. Johnston, eds. *The Middling Sorts: Explorations in the History of the American Middle Class.* New York and London: Routledge Press, 2000.

Bloch, Ruth H. *Gender and Morality in Anglo-American Culture, 1650–1800.* Berkeley, Los Angeles, and London: University of California Press, 2003.

————. "The Gendered Meanings of Virtue in Revolutionary America." *Signs: Journal of Women in Culture and Society* 13 (Autumn 1987): 37–58.

————. "American Feminine Ideals in Transition: The Rise of the Moral Mother, 1785–1815." *Feminist Studies* 4 (June 1978): 101–126.

Blumin, Stuart M., Robert Fogel, and Stephan Thernstrom, eds. *The Emergence of the Middle Class: Social Experience in the American City, 1760–1900.* Cambridge: Cambridge University Press, 1989.

Borer, Mary. *Willingly to School: A History of Women's Education.* Guildford: Lutterworth Press, 1976.

Boydston, Jeanne. *Home & Work: Housework, Wages, and the Ideology of Labor in the Early Republic.* New York and Oxford: Oxford University Press, 1990.

————, Mary Kelley, and Anne Margolis. *The Limits of Sisterhood: The Beecher Sisters on Women's Rights and Woman's Sphere.* Chapel Hill and London: The University of North Carolina Press, 1988.

Boylan, Anne M. *The Origins of Women's Activism: New York and Boston, 1797–1840.* Chapel Hill: The University of North Carolina Press, 2002.

Brekus, Catherine A. *Strangers & Pilgrims: Female Preaching in America, 1740–1845.* Chapel Hill: The University of North Carolina Press, 1998.

Brenzel, Barbara M. "History of 19th Century Women's Education: A Plea for Inclusion of Class, Race and Ethnicity." Working Paper No. 114, Wellesley College Center for Research on Women, 1983.

Brickley, Lynn Templeton. "Sarah Pierce's Litchfield Female Academy, 1792–1833." Ph.D. Dissertation, Harvard University, 1985.

Broome, Edwin C. *A Historical and Critical Discussion of College Admission Requirements.* New York: Macmillan & Co., 1903.

Brown, Richard D. *Knowledge Is Power: The Diffusion of Information in Early America, 1700–1965.* New York and Oxford: Oxford University Press, 1989.

Bushman, Richard L. *The Refinement of America: Persons, Houses, Cities.* New York: Random House, 1992.

Butler, Judith. *Gender Trouble: Feminism and the Subversion of Identity*. New York: Routledge, 1990.

Carnes, Mark C. *Secret Ritual and Manhood in Victorian America*. New Haven, CT: Yale University Press, 1989.

Casement, William. "Learning and Pleasure: Early American Perspectives." *Educational Theory* 40 (Summer 1990): 343–349.

Chambers-Schiller, Lee Virginia. *Liberty, A Better Husband: Single Women in America: The Generations of 1780–1840*. New Haven, CT and London: Yale University Press, 1984.

Church, Robert L. *Education in the United States: An Interpretive History*. New York: The Free Press, 1976.

Clifford, Geraldine Jonich. "Man/Woman/Teacher: Gender, Family and Career in American Educational History." In *American Teachers: Histories of a Profession at Work*, Donald Warren, ed. New York: Macmillan Publishing Company, 1989, 293–343.

Clinton, Catherine. "Equally Their Due: The Education of the Planter Daughter in the Early Republic." *Journal of the Early Republic* (April 1982): 39–60.

Cogan, Frances B. *All-American Girl: The Ideal of Real Womanhood in Mid-Nineteenth Century America*. Athens: University of Georgia Press, 1989.

Cohen, Daniel A. "The Respectability of Rebecca Reed: Genteel Womanhood and Sectarian Conflict in Antebellum America." *Journal of the Early Republic* 16 (Fall 1996): 419–461.

Cohen, I. Bernard. *Science and the Founding Fathers: Science in the Political Thought of Jefferson, Franklin, Adams, and Madison*. New York and London: W. W. Norton & Co., 1995.

Conforti, Joseph. *Jonathan Edwards, Religious Tradition, and American Culture*. Chapel Hill and London: The University of North Carolina Press, 1995.

———. "Mary Lyon, the Founding of Mount Holyoke College, and the Cultural Revival of Jonathan Edwards." *Religion and American Culture: A Journal of Interpretation* 3 (Winter 1993): 69–89.

Conway, Jill K. "Perspectives on the History of Women's Education in the United States." *History of Education Quarterly* 14 (Spring 1974): 1–12.

Cott, Nancy. *The Bonds of Womanhood: "Women's Sphere" in New England, 1780–1835*. New Haven, CT and London: Yale University Press, 1977.

Coulter, E. Merton. "The Ante-Bellum Academy Movement in Georgia." *Georgia Historical Quarterly* V (December 1921): 11–42.

Cremin, Lawrence. *American Education: The National Experience, 1783–1876*. New York: Harper & Row, 1980.

Davidson, Cathy N. "No More Separate Spheres!" *American Literature* 70 (September 1998): 443–463.

———. "Female Education, Literacy and the Politics of Sentimental Fiction." *Women's Studies International Forum* 9 (1986): 309–312.

Douglas, Ann. *The Feminization of American Culture*. New York: Doubleday, 1977.

Dowling, William. *Literary Federalism in the Age of Jefferson*. Columbia: University of South Carolina Press, 1999.

Drakeman, Lisa Natale. "Seminary Sisters: Mount Holyoke's First Students, 1837–1849." Ph.D. Dissertation, Princeton University, 1988.

Dubbert, Joe L. "Progressivism and the Masculinity Crisis." In *The American Man*, Elizabeth H. Pleck and Joseph H. Pleck, eds. Englewood Cliffs, NJ: Prentice-Hall, 1980, 303–320.

Dublin, Thomas, ed. *Farm to Factory: Women's Letters, 1830–1860*. New York: Columbia University Press, 1981.

DuBois, Ellen Carol, ed. *The Elizabeth Cady Stanton—Susan B. Anthony Reader: Correspondence, Writings, Speeches*. Boston: Northeastern University Press, 1992.

————. *Feminism and Suffrage: The Emergence of an Independent Women's Movement in America, 1848–1869*. Ithaca, NY: Cornell University Press, 1978.

Eisenmann, Linda. "Reconsidering a Classic: Assessing the History of Women's Higher Education a Dozen Years after Barbara Solomon." *Harvard Educational Review* 67 (Winter 1997): 689–717.

Elsbree, Willard S. *The American Teacher: Evolution of a Profession in a Democracy*. New York: American Book Company, 1939.

Epstein, Barbara Leslie. *The Politics of Domesticity: Women, Evangelism, and Temperance in Nineteenth Century America*. Middletown, CT: Wesleyan University Press, 1981.

Evans, Sara M. *Born for Liberty: A History of Women in America*. New York: The Free Press, 1989.

Farnham, Christie Anne. *The Education of the Southern Belle: Higher Education and Student Socialization in the Antebellum South*. New York and London: New York University Press, 1994.

Fatherly, Sarah E. "Gentlewomen and Learned Ladies: Gender and the Creation of an Urban Elite in Colonial Philadelphia." Ph.D. Dissertation, University of Wisconsin-Madison, 2000.

Filene, Peter G. *Him/Her/Self: Sex Roles in Modern America*. Baltimore: The Johns Hopkins University Press, 1986.

Flexner, Eleanor. *Century of Struggle: The Woman's Rights Movement in the United States*. Cambridge: The Belknap Press of Harvard University Press, 1959.

Foner, Philip S. and Josephine F. Pacheco. *Three Who Dared: Prudence Crandall, Margaret Douglass, Myrtilla Miner: Champions of Antebellum Black Education*. Westport, CT: Greenwood Press, 1984.

Foucault, Michel. *The History of Sexuality*, vol. 1. New York: Vintage Books, 1978.

Fowler, Elaine W. *Life in the New Nation, 1787–1860*. New York: Capricorn Books, 1974.

Frankfurt, Roberta. *Collegiate Women: Domesticity and Career in Turn-of-the-Century America*. New York: New York University Press, 1977.

Fraser, Nancy. *Unruly Practices: Power, Discourse, and Gender in Contemporary Social Theory*. Minneapolis: University of Minnesota Press, 1989.

Geiger, Roger "The Superior Education of Women." *The Transformation of the Colleges*, Unpublished draft.

Geiger, Roger L. "The Historical Matrix of American Higher Education." *History of Higher Education Annual* 12 (1992): 7–28.

Gelles, Edith B. *Portia: The World of Abigail Adams*. Bloomington: Indiana University Press, 1992.

Ginzberg, Lori D. *Women and the Work of Benevolence: Morality, Politics, and the Class in the 19th-Century United States*. New Haven, CT and London: Yale University Press, 1990.

Goodsell, Willystine. *The Education of Women: Its Social Background and Its Problems*. New York: Macmillan Co., 1923.

Goodwin, Joan W. *The Remarkable Mrs. Ripley: The Life of Sarah Alden Bradford Ripley*. Boston: Northeastern University Press, 1998.

Gordon, Ann. "The Young Ladies Academy of Philadelphia." In *Women of America: A History*, Carol Ruth Berkin and Mary Beth Norton, eds. Boston: Houghton Mifflin, 1979: 68–91.

Gordon, Lynn D. *Gender and Higher Education in the Progressive Era*. New Haven, CT and London: Yale University Press, 1990.

Gordon, Sarah. "Smith College Students: The First Ten Classes, 1879–1888." *History of Education Quarterly* 15 (Summer 1975): 147–167.

Graham, Patricia Albjerg. "Expansion and Exclusion: A History of Women in American Higher Education." *Signs* 3 (Summer 1978): 759–773.

Green, Elizabeth Alden. *Mary Lyon and Mount Holyoke: Opening the Gates*. Hanover, NH: University Press of New England, 1979.

Green, Harvey. *Fit for America: Health, Fitness, Sport, and American Society*. Baltimore: The Johns Hopkins University Press, 1986.

Guralnik, Stanley, M. *Science and the Ante-Bellum American College*. Philadelphia: American Philosophical Society, 1975.

Habermas, Jurgen. *The Structural Transformation of the Public Sphere: An Inquiry into a Category of Bourgeois Society* [1962], trans. Thomas Burger. Cambridge: MIT Press, 1994.

Handler, Bonnie. "Prudence Crandall and Her School for Young Ladies and Little Misses of Color." *Vitae Scholasticae* 5 (Winter 1986): 199–210.

Hansen, Karen. *A Very Social Time: Crafting Community in Antebellum New England*. Berkeley: University of California Press, 1994.

Hantover, Jeffrey P. "The Boy Scouts and the Validation of Masculinity." In *The American Man*, Elizabeth H. Pleck and Joseph H. Pleck, eds. Englewood Cliffs, NJ: Prentice-Hall, 1980: 285–302.

Herbert, T. Walter. *Dearest Beloved: The Hawthornes and the Making of the Middle-Class Family*. Berkeley: University of California Press, 1993.

Herbst, Jurgen. *And Sadly Teach: Teacher Education and Professionalization in American Culture*. Madison: University of Wisconsin Press, 1989.

Hewitt, Nancy A. "Taking the True Woman Hostage." *Journal of Women's History* 14 (Spring 2002): 156–162.

———. *Women's Activism and Social Change: Rochester, New York, 1822–1872*. Ithaca, NY and London: Cornell University Press, 1984.

Higham, John, ed. "The Reorientation of American Culture in the 1890s." *Writing American History: Essays on Modern Scholarship*. Bloomington: Indiana University Press, 1978, 73–102.

Hine, Darlene Clark, Elsa Barkley Brown, and Rosalyn Terborg-Penn. *Black Women in America: An Historical Encyclopedia.* Bloomington: Indiana University Press, 1993.

Hoeveler, J. David. *Creating the American Mind: Intellect and Politics in the Colonial Colleges.* Lanham, Boulder, New York and Oxford: Rowman & Littlefield, 2002.

Holmes, Madelyn and Beverly J. Weiss. *Lives of Women Public Schoolteachers: Scenes from American Educational History.* New York and London: Garland Publishing, Inc., 1995.

Horowitz, Helen Lefkowitz. *Alma Mater: Design and Experience in Women's Colleges from Their Nineteenth-Century Beginnings to the 1930s.* New York: Knopf, 1984.

Horton, James Oliver and Lois E. Horton. *In Hope of Liberty: Culture, Community, and Protest among Northern Free Blacks, 1700–1860.* New York: Oxford University Press, 1997.

Howe, Florence. "Myths of Coeducation." *Myths of Coeducation—Selected Essays, 1964–1983.* Bloomington: Indiana University Press, 1984: 206–220.

Jabour, Anya. *Marriage in the Early Republic: Elizabeth and William Wirt and the Companionate Ideal.* Baltimore and London: The Johns Hopkins University Press, 1998.

———. " 'Grown Girls, Highly Cultivated': Female Education in an Antebellum Southern Family." *Journal of Southern History* LXIV (February 1998): 23–64.

Jeffrey, Julie Roy. "Permeable Boundaries: Abolitionist Women and Separate Spheres." *Journal of the Early Republic* 21 (Spring 2001): 79–93.

———. *The Great Silent Army of Abolitionism: Ordinary Women in the Antislavery Movement.* Chapel Hill and London: The University of North Carolina Press, 1998.

Johnson, Clifton. *Old-Time Schools and School-Books.* Gloucester, MA: Peter Smith, 1963.

Johnson, Paul E. *A Shopkeeper's Millennium: Society and Revivals in Rochester, New York, 1815–1837.* New York: Hill and Wang, 1978.

Kaestle, Carl F. *Pillars of the Republic: Common Schools and American Society, 1780–1860.* New York: Hill and Wang, 1983.

Katz, Michael B. *The Social Organization of Early Industrial Capitalism.* Cambridge: Cambridge University Press, 1982.

Kelley, Mary. " 'A More Glorious Revolution': Women's Antebellum Reading Circles and the Pursuit of Public Influence." *New England Quarterly* 76 (June 2003): 163–196.

———. "Beyond the Boundaries." *Journal of the Early Republic* 21 (Spring 2001): 73–78.

———. "Reading Women/Women Reading: The Making of Learned Women in Antebellum America." *Journal of American History* 83 (September 1996): 401–424.

———. *Private Woman Public Stage: Literary Domesticity in Nineteenth-Century America.* New York: Oxford University Press, 1984.

———. "The Sentimentalists: Promise and Betrayal in the Home." *Signs* 4 (Spring 1979): 434–446.

Kendall, Elaine. *Peculiar Institutions: An Informal History of the Seven Sister Colleges.* New York: Putnam's, 1975.

Kerber, Linda K. *No Constitutional Right to be Ladies: Women and the Obligations of Citizenship*. New York: Hill and Wang, 1998.

———. "Beyond Roles, Beyond Spheres: Thinking about Gender in the Early Republic." *William and Mary Quarterly* 3rd ser., 44 (July 1989): 565–581; XLVI (July 1989): 577; 46 (July 1989): 565–585.

———. *Women of the Republic: Intellect & Ideology in Revolutionary America*. Chapel Hill: The University of North Carolina Press, 1980.

———. "The Republican Mother: Women and the Enlightenment—An American Perspective." *American Quarterly* 28 (Summer 1976): 187–205.

———. "Daughters of Columbia: Educating Women for the Republic, 1787–1805." In *The Hofstader Aegis: A Memorial*, Stanley Elkins and Eric McKitrick, eds. New York: Alfred A. Knopf, 1974, 36–59.

——— and Jane DeHart-Mathews, eds. *Women's America: Refocusing the Past*. New York and Oxford: Oxford University Press, 1987.

Kerns, Kathryn M. "Farmers' Daughters: The Education of Women at Alfred Academy and University Before the Civil War." *History of Higher Education Annual* 6 (1986): 10–28.

Kerr, Andrea Moore. *Lucy Stone: Speaking Out for Equality*. New Brunswick, NJ: Rutgers University Press, 1992.

Kett, Joseph F. *The Pursuit of Knowledge Under Difficulties: From Self-Improvement to Adult Education in America, 1750–1990*. Stanford, CA: Stanford University Press, 1994.

Kierner, Cynthia A. *Beyond the Household: Women's Place in the Early South, 1700–1835*. Ithaca, NY and London: Cornell University Press, 1998.

Kimmel, Michael S. "The Contemporary 'Crisis' of Masculinity in Historical Perspective." In *The Making of Masculinities*, Harry Brod, ed. Boston: Allen & Unwin, 1987: 121–154.

Kinnard, Cynthia. *Antifeminism in American Thought: An Annotated Bibliography*. Boston: G. K. Hall, 1986.

Laqueur, Thomas. *Making Sex: Body and Gender from the Greeks to Freud*. Cambridge, MA and London: Harvard University Press, 1992.

Lasser, Carol. "Beyond Separate Spheres: The Power of Public Opinion." *Journal of the Early Republic* 21 (Spring 2001): 115–123.

Lawson, Ellen N. and Marlene Merrill. "The Antebellum 'Talented Thousandth': Black College Students at Oberlin Before the Civil War." *Journal of Negro Education* 52 (Spring 1983): 142–155.

Lehmann, William C. *John Millar of Glasgow, 1735–1801*. London and New York: Cambridge University Press, 1960.

Lerner, Gerda. *The Woman in American History*. Menlo Park: Addison-Wesley Publishing Co., 1971.

———. "The Lady and the Mill Girl: Changes in the Status of Women in the Age of Jackson." *Midcontinent American Studies Journal* 10 (Winter 1969): 5–15.

Lewis, Jan. "The Republican Wife: Virtue and Seduction in the Early Republic." *William and Mary Quarterly* 3rd ser., XLIV (October 1987): 689–721.

———. *The Pursuit of Happiness: Family and Values in Jefferson's Virginia*. New York: Cambridge University Press, 1983.

Lindman, Janet Moore. "Acting the Manly Christian: White Evangelical Masculinity in Revolutionary Virginia." *William and Mary Quarterly* 3rd ser., LVII (April 2000), 393–416.

Lockridge, Kenneth A. *Literacy in Colonial New England: An Enquiry into the Social Context of Literacy in the Early Modern West.* New York: W. W. Norton & Co., 1974.

Lyman, Rollo Laverne. *English Grammar in American Schools Before 1850.* Washington: Government Printing Office, 1922.

Main, Gloria L. "An Inquiry into When and Why Women Learned to Write in Colonial New England." *Journal of Social History* 24 (Spring 1991): 579–589.

Matthews, Glenna. *The Rise of Public Woman: Woman's Power and Woman's Place in the United States, 1630–1970.* New York: Oxford University Press, 1992.

Mattingly, Paul H. *The Classless Profession: American Schoolmen in the Nineteenth Century.* New York: New York University Press, 1975.

May, Henry F. *The Enlightenment in America.* New York: Oxford University Press, 1976.

McCall, Laura. " 'Shall I Fetter Her Will?': Literary Americans Confront Feminine Submission, 1820–1860." *Journal of the Early Republic* 21 (Spring 2001): 95–113.

————. " 'The Reign of Brute Force Is Now Over': A Content Analysis of *Godey's Lady's Book*, 1830–1860." *Journal of the Early Republic* 9 (Summer 1989): 217–236.

McClelland, Clarence P. *The Education of Females in Early Illinois.* Jacksonville, IL: MacMurray College for Women, 1944.

McGovern, James R. "David Graham Phillips and the Virility Impulse of the Progressives." *New England Quarterly* 39 (September 1966): 334–355.

McLoughlin, William G. *Revivals, Awakenings, and Reform: An Essay on Religion and Social Change in America, 1607–1977.* Chicago: University of Chicago Press, 1978.

Melder, Keith E. "Masks of Oppression: The Female Seminary Movement in the United States." In *History of Women in the United States: Historical Articles on Women's Lives and Activities*, vol. 12, Nancy F. Cott, ed. Munich, New Providence, London, and Paris: K. G. Saur, 1993, 25–44.

————. *Beginnings of Sisterhood: The American Woman's Rights Movement, 1800–1850.* New York: Schocken Books, 1977.

Miller, George Frederick. *The Academy System of the State of New York.* New York: Arno Press & The New York Times, 1969.

Millis, William Alfred. *The History of Hanover College from 1827 to 1927.* Hanover, NH: Hanover College, 1927.

Morison, Samuel Eliot. *Three Centuries of Harvard, 1636–1936.* Cambridge: The Belknap Press of Harvard University Press, 1964.

Naeve, Milo M. *John Lewis Krimmel: An Artist in Federal America.* Newark: University of Delaware Press, 1987.

Nash, Margaret A. "Rethinking Republican Motherhood: Benjamin Rush and the Young Ladies' Academy of Philadelphia." *Journal of the Early Republic* 17 (Summer 1997): 171–192.

Newcomer, Mabel. *A Century of Higher Education for American Women*. New York: Harper & Row, 1959.

Nichols, Heidi L. *The Fashioning of Middle-Class America: Sartains Union Magazine of Literature and Art and Antebellum Culture*. New York: Peter Lang, 2004.

Nietz, John A. *Old Textbooks*. Pittsburgh: University of Pittsburgh Press, 1961.

Norton, Mary Beth. *Liberty's Daughters: The Revolutionary Experience of American Women, 1750–1800*. New York: HarperCollins, 1980.

Nye, Russel Blaine. *The Cultural Life of the New Nation, 1776–1830*. New York and Evanston: Harper & Row, 1960.

Oates, Mary J. and Susan Williamson. "Women's Colleges and Women Achievers." *Signs* 3 (Summer 1978): 795–806.

Ogren, Christine A. "Where Coeds were Coeducated: Normal Schools in Wisconsin, 1870–1920." *History of Education Quarterly* 35 (Spring 1995): 1–26.

Palmieri, Patricia A. *In Adamless Eden*. New Haven, CT and London: Yale University Press, 1995.

———. "Here was Fellowship: A Social Portrait of Academic Women at Wellesley College, 1880–1920." *History of Education Quarterly* 23 (Summer 1983): 195–214.

———. "Patterns of Achievement of Single Academic Women at Wellesley College, 1880–1920." *Frontiers* 5 (Spring 1980): 63–67.

Pangle, Lorraine Smith and Thomas L. Pangle. *The Learning of Liberty*. Lawrence: University Press of Kansas, 1993.

Pease, Jane H. and William H. Pease. *Ladies, Women & Wenches: Choice and Constraint in Antebellum Charleston and Boston*. Chapel Hill: University of North Carolina Press, 1990.

Perkins, Linda M. "Heed Life's Demands: The Educational Philosophy of Fanny Jackson Coppin." *Journal of Negro Education* 51 (Summer 1982): 181–190.

Perlman, Joel and Robert A. Margo. *Women's Work? American Schoolteachers, 1650–1920*. Chicago: University of Chicago, 2001.

Perry, Lewis. *Intellectual Life in America*. New York: Franklin Watts, 1984.

Perry, Ruth. "Mary Astell's Response to the Enlightenment." *Women and History* 9 (Spring 1984): 13–40.

Pond, Jean. *Bradford: A New England Academy*. Bradford, MA: Alumnae Association, 1930.

Porter, Dorothy B. "The Organized Educational Activities of Negro Literary Societies, 1828–1846." *Journal of Negro Education* 5 (October 1936): 555–576.

Potter, Robert E. *The Stream of American Education*. New York: American Book Company, 1967.

Preston, JoAnne. "Domestic Ideology, School Reformers, and Female Teachers: Schoolteaching Becomes Women's Work in Nineteenth-Century New England." *The New England Quarterly* LXVI (December 1993): 531–551.

Rammelkamp, Charles Henry. *Illinois College: A Centennial History, 1829–1929*. New Haven, CT: Yale University Press, 1928.

Reese, William J. *The Origins of the American High School*. New Haven, CT and London: Yale University Press, 1995.

Reinier, Jacqueline S. "Rearing the Republican Child: Attitudes and Practices in Post-Revolutionary Philadelphia." *William and Mary Quarterly* 39 (January 1982): 150–163.

Richard, Carl J. *The Founders and the Classics: Greece, Rome, and the American Enlightenment.* Cambridge: Harvard University Press, 1994.

Richardson, Marilyn, ed. *Maria W. Stewart, America's First Black Woman Political Writer: Essays and Speeches.* Bloomington: Indiana University Press, 1987.

Riley, Glenda. "Origins of the Argument for Improved Female Education." *History of Education Quarterly* 9 (Winter 1969): 455–470.

Robson, David W. *Educating Republicans: The College in the Era of the American Revolution, 1750–1800.* Westport, CT and London: Greenwood Press, 1985.

Rosenberg, Rosalind. *Beyond Separate Spheres: Intellectual Roots of Modern Feminism.* New Haven, CT: Yale University Press, 1982.

Rossiter, Margaret W. *Women Scientists in America: Struggles and Strategies to 1940.* Baltimore: The Johns Hopkins University Press, 1982.

Rotundo, E. Anthony. *American Manhood: Transformations in Masculinity from the Revolution to the Modern Era.* New York: Basic Books, 1993.

Rudolph, Frederick. *The American College and University: A History.* New York: Alfred A. Knopf, 1968.

Rury, John L. "The Era of Republican Motherhood: A Formative Period in the History of American Women's Education." *Journal of the Midwest History of Education Society* 20 (1992): 151–170.

———. "Who Became Teachers and Why: The Social Characteristics of Teachers in American History." In *American Teachers: Histories of a Profession at Work,* Donald Warren, ed. New York: Macmillan Publishing Company, 1989: 9–48.

Ryan, Mary P. *Women in Public: Between Banners and Ballots, 1825–1880.* Baltimore and London: The Johns Hopkins University Press, 1990.

———. *The Cradle of the Middle-Class: The Family in Oneida County, New York, 1780–1865.* Cambridge: Cambridge University Press, 1981.

Schiebinger, Londa. *The Mind Has No Sex? Women in the Origins of Modern Science.* Cambridge and London: Harvard University Press, 1989.

Schloesser, Pauline. *The Fair Sex: White Women and Racial Patriarchy in the Early American Republic.* New York and London: New York University Press, 2002.

Scott, Anne Firor. "The Ever-Widening Circle: The Diffusion of Feminist Values from the Troy Female Seminary, 1822–1872." *History of Education Quarterly* 19 (Spring 1979): 3–24.

Sellers, Charles. *The Market Revolution: Jacksonian America, 1815–1846.* New York: Oxford University Press, 1991.

Shalhope, Robert E. "Toward a Republican Synthesis: The Emergence of an Understanding of Republicanism in American Historiography." *William and Mary Quarterly* 3rd ser., 29 (January 1972): 49–80.

Shoemaker, Robert B. *Gender in English Society, 1650–1850: The Emergence of Separate Spheres?* London and New York: Longman, 1998.

Sizer, Theodore R., ed. *The Age of the Academies.* New York: Teachers College, 1964.

Sklar, Kathryn Kish. "The Founding of Mount Holyoke College." In *Women of America: A History*, Carol Ruth Berkin and Mary Beth Norton, eds. Boston: Houghton Mifflin Co., 1979, 177–201.
———. *Catharine Beecher: A Study in American Domesticity*. New York: W. W. Norton & Co., 1976.
Smith, Charles Lee. *The History of Education in North Carolina*. Washington: Government Printing Office, 1888.
Smith, Daniel Blake. *Inside the Great House: Planter Family Life in Eighteenth-Century Chesapeake Society*. Ithaca, NY: Cornell University Press, 1980.
Smith, Daniel Scott. "Parental Power and Marriage Patterns: An Analysis of Historical Trends in Hingham, Massachusetts." *Journal of Marriage and the Family* 35 (August 1973): 419–428.
Smith-Rosenberg, Carroll. *Religion and the Rise of the American City: The New York City Mission Movement, 1812–1820*. Ithaca, NY: Cornell University Press, 1971.
———. "Beauty, the Beast and the Militant Woman: A Case Study in Sex Roles and Social Stress in Jacksonian America." *American Quarterly* 23 (October 1971): 562–584.
Solomon, Barbara Miller. *In the Company of Educated Women: A History of Women and Higher Education in America*. New Haven, CT and London: Yale University Press, 1985.
Soltow, Lee and Edward Stevens. *The Rise of Literacy and the Common School in the United States: A Socioeconomic Analysis to 1870*. Chicago: University of Chicago Press, 1981.
Stansell, Christine. *City of Women: Sex and Class in New York, 1789–1860*. Urbana and Chicago: University of Illinois Press, 1987.
Sterling, Dorothy. *Ahead of Her Time: Abby Kelley and the Politics of Antislavery*. New York and London: W. W. Norton & Co., 1991.
———. *We Are Your Sisters: Black Women in the Nineteenth Century*. New York and London: W. W. Norton & Co., 1984.
Sumler-Edmond, Janice. "Charlotte L. Forten Grimke." In *Black Women in America: An Historical Encyclopedia*, Darlene Clark Hine et al., eds. Bloomington: Indiana University Press, 1993: 505.
Swan, Susan Burrows. *Plain and Fancy: American Women and Their Needlework, 1700–1850*. New York: Routledge, 1977.
Teaford, John. "The Transformation of Massachusetts Education, 1670–1780." In *The Social History of American Education*, B. Edward McClellan and William J. Reese, eds. Urbana and Chicago: University of Illinois Press, 1988, 23–38.
Tewksbury, Donald G. *The Founding of American Colleges and Universities Before the Civil War*. New York: Arno Press & The New York Times, 1969.
Todd, Jan. *Physical Culture and the Body Beautiful: Purposive Exercise in the Lives of American Women, 1800–1870*. Macon, GA: Mercer University Press, 1998.
Tolley, Kim. *The Science Education of American Girls: A Historical Perspective*. New York and London: RoutledgeFalmer, 2003.
———. "Mapping the Landscape of Higher Schooling, 1727–1850." In *Chartered Schools: Two Hundred Years of Independent Academies in the United States,*

*1727–1925*, Nancy Beadie and Kim Tolley, eds. New York and London: RoutledgeFalmer, 2002, 19–43.

Townsend, Lucy Forsyth. "Emma Willard: Eclipse or Reemergence?" *Journal of the Midwest History of Education Society* 18 (1990): 279–292.

Townsend, Lucy F. and Barbara Wiley. "Divorce and Domestic Education: The Case of Emma Willard." Paper Presented at the History of Education Society Annual Meeting, October, 1993.

Trachtenberg, Alan. *The Incorporation of America: Culture and Politics in the Gilded Age.* New York: Hill and Wang, 1982.

Tuckerman, Frederick. *Amherst Academy: A New England School of the Past, 1814–1861.* Amherst: Trustees, 1929.

Tyack, David and Elisabeth Hansot. *Learning Together: A History of Coeducation in American Schools.* New Haven, CT and London: Yale University Press, 1990.

———. *Managers of Virtue: Public School Leadership in America, 1820–1980.* New York: Basic Books, 1982.

Tyack, David and Myra H. Strober. "Jobs and Gender: A History of the Structuring of Educational Employment by Sex." In *Educational Policy and Management: Sex Differentials*, Patricia Schmuck, ed. New York: Academic Press, 1981.

Vanderpoel, Emily. *Chronicles of a Pioneer School, from 1792 to 1833.* Cambridge, MA: Cambridge University Press, 1903.

———. *More Chronicles of a Pioneer School, from 1792 to 1833.* New York: The Cadmus Book Shop, 1927.

Varon, Elizabeth R. *We Mean to be Counted: White Women & Politics in Antebellum Virginia.* Chapel Hill and London: The University of North Carolina Press, 1998.

Vinovskis, Maris A. and Richard M. Bernard. "Beyond Catharine Beecher: Female Education in the Antebellum Period." *Signs: Journal of Women in Culture and Society* 3 (Summer 1978): 856–869.

Walsh, James J. *Education of the Founding Fathers of the Republic: Scholasticism in the Colonial Colleges.* New York: Fordham University Press, 1935.

Warner, Deborah Jean. "Science Education for Women in Antebellum America." *Isis* 69 (March 1978): 58–67.

Warren, Donald, ed. *American Teachers: Histories of a Profession at Work.* New York: Macmillan, 1989.

Washburn, Emory. *Brief Sketch of a History of the Leicester Academy.* Boston: Phillips, Sampson, 1855.

Watkinson, James D. "Useful Knowledge? Concepts, Values, and Access in American Education, 1776–1840." *History of Education Quarterly* 30 (Fall 1990): 351–370.

Watson, Harry L. *Liberty and Power: The Politics of Jacksonian America.* New York: Hill and Wang, 1990.

Wells, Jonathan Daniel. *The Origins of the Southern Middle Class, 1800–1861.* Chapel Hill: The University of North Carolina Press, 2004.

Welter, Barbara. "The Cult of True Womanhood." *American Quarterly* 18 (Summer 1966): 151–174.

Wheeler, Kenneth H. "Why the Early Coeducational College Was Primarily a Midwestern Phenomenon." Paper Presented at the History of Education Society Annual Meeting, October 30, 1998.

White, Marian Churchill. *A History of Barnard College.* New York: Columbia University Press, 1954.

Winch, Julie. " 'You Have Talents—Only Cultivate Them': Philadelphia's Black Female Literary Societies and the Abolitionist Crusade." In *The Abolitionist Sisterhood: Women's Political Culture in Antebellum America*, Jean Fagan Yellin and John C. Van Home, eds. Ithaca, NY: Cornell University Press, 1994: 101–118.

Winterer, Caroline. *The Culture of Classicism: Ancient Greece and Rome in American Intellectual Life, 1780–1910.* Baltimore and London: The Johns Hopkins University Press, 2002.

Woody, Thomas. *A History of Women's Education in the United States.* New York: The Science Press, 1929.

Wright, Louis B. and Elaine W. Fowler. *Life in the New Nation, 1787–1860.* New York: Capricorn Books, 1974.

Yee, Shirley J. *Black Women Abolitionists: A Study in Activism, 1828–1860.* Knoxville: University of Tennessee Press, 1992.

Yellin, Jean Fagan and John C. Van Home, eds. *The Abolitionist Sisterhood: Women's Political Culture in Antebellum America.* Ithaca, NY: Cornell University Press, 1994.

Young, Iris. *Justice and the Politics of Difference.* Princeton, NY: Princeton University Press, 1990.

Young, Linda. *Middle Class Culture in the Nineteenth Century: America, Australia and Britain.* New York: Palgrave Macmillan, 2003.

Zaeske, Susan. *Signatures of Citizenship: Petitioning, Antislavery, and Women's Political Identity.* Chapel Hill and London: The University of North Carolina Press, 2003.

Zagarri, Rosemarie. "Morals, Manners, and the Republican Mother." *American Quarterly* 44 (June 1992): 192–215.

Zboray, Ronald J. *A Fictive People: Antebellum Economic Development and the American Reading Public.* New York: Oxford University Press, 1993.

———, and Mary Saracino Zboray. " 'Months of Mondays': Women's Reading Diaries and the Everyday Transcendental." Paper Presented at the Eleventh Berkshire Conference on the History of Women, June 6, 1999.

# Index

CPSIA information can be obtained at www.ICGtesting.com
Printed in the USA
LVOW010046211212

312712LV00008B/149/P